Incomplete Decompositions (ILU) –
Algorithms, Theory, and Applications

Edited by
Wolfgang Hackbusch and
Gabriel Wittum

Notes on Numerical Fluid Mechanics (NNFM) Volume 41

Incomplete Decomposition (ILU) – Algorithms, Theory, and Applications

Proceedings of the Eighth GAMM-Seminar,
Kiel, January 24–26, 1992

Edited by
Wolfgang Hackbusch and
Gabriel Wittum

Mathematics Subject Classification: 76-XX

Produced by W. Langelüddecke, Braunschweig
Printed on acid-free paper

ISSN 0179-9614
ISBN 978-3-528-07641-2   ISBN 978-3-322-85732-3 (eBook)
DOI 10.1007/978-3-322-85732-3

Preface

The GAMM Committee for "Efficient Numerical Methods for Partial Differential Equations" organizes workshops on subjects concerning the algorithmical treatmant of partial differential equations. The topics are discretization methods like the finite element and finite volume method for various types of applications in structural and fluid mechanics. Particular attention is devoted to advanced solution techniques.

The series of such workshops was continued in 1992, January 24-26, with the 8th Kiel-Seminar on the special topic

"Incomplete Decompositions (ILU)
Algorithms, Theory and Applications"

at the Christian-Albrechts-University of Kiel. The seminar was attended by 70 scientists from 9 countries and 22 lectures were given.

The list of topics contained applications of incomplete decompositions as smoothers in multi-grid methods, as preconditioners for conjugate gradient type methods and the use of ILU for systems of partial differential equations.

This volume is dedicated to the memory of Werner Weiler, who compiled a large part of the papers herein. He died tragically in spring 1992.

September 1992

Wolfgang Hackbusch
Gabriel Wittum

Contents

RELAXED ILU PRECONDITIONING FOR THE CG SOLUTION OF
A SINGULAR BOUNDARY VALUE PROBLEM FOR PRESSURE

Vincent A. Barker
Institute for Numerical Analysis
Technical University of Denmark
DK-2800 Lyngby, Denmark

Bo Gervang *
Department of Fluid Mechanics
Technical University of Denmark
DK-2800 Lyngby, Denmark

Summary

A splitting technique for solving the incompressible Navier-Stokes equations, whereby operations on pressure and velocity are decoupled, leads to a second-order singular boundary value problem for the determination of pressure. A finite volume discretization based on a nonuniform grid is applied to this problem, producing a linear system of equations whose coefficient matrix is symmetric positive semidefinite. We give an account of our experience with solving this system by the conjugate gradient method with relaxed ILU preconditioning. Particular attention is given to the effect of the nonuniformity of the grid.

1. A singular boundary value problem and its discretization

A splitting technique for solving the Navier-Stokes equations, whereby operations on pressure and velocity are decoupled, leads to the singular boundary value problem

$$-p_{xx} -p_{yy} = f(x,y), \quad (x,y) \in \Omega \tag{1a}$$

$$p_n = g(x,y), \quad (x,y) \in \partial\Omega \tag{1b}$$

* Present address: Department of Mathematics, University College of Wales, Aberystwyth SY23 3BZ, UK.

This work has been supported in part by the Danish Natural Science Research Council.

1

where Ω is the open rectangle

$$\Omega = \left\{ (x,y) \in R^2 \mid 0 < x < L_1,\ 0 < y < L_2 \right\} .$$

The functions f and g are smooth and satisfy the consistency condition

$$\iint_\Omega f\,dxdy + \int_{\partial\Omega} g\,ds = 0 , \tag{2}$$

so problem (1) has an infinite number of solutions. If p^* is any particular solution then the set of solutions is $p = p^*+c$ for every constant c .

We apply the finite volume method to (1), using a nonuniform grid whose parameters are positive integers m and n and positive real constants α and β. (See Figure 1). The grid points, denoted $(x_j,\ y_k)$ for $j = 0,1,2,\ldots,m$ and $k = 0,1,2,\ldots,n,$ are determined by the conditions

$$x_{j+1} - x_j = \alpha \left(x_j - x_{j-1}\right),\ j = 1,2,\ldots,m-1 \tag{3a}$$
$$y_{k+1} - y_k = \beta \left(y_k - y_{k-1}\right),\ k = 1,2,\ldots,n-1 . \tag{3b}$$

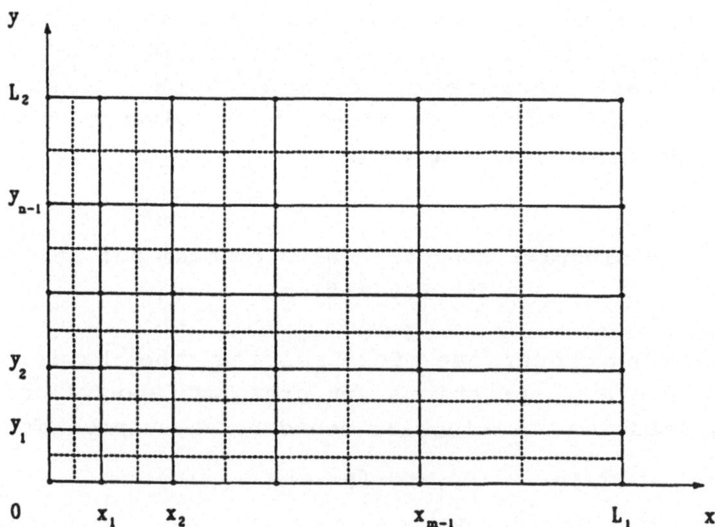

Fig. 1. A nonuniform grid for the finite volume method.

If (as is the case in our applications) $\alpha > 1$ and $\beta > 1$ then there is a concentration of grid points at the origin, the concentration becoming more pronounced as α and β increase. It is readily seen that

$$x_1 = \frac{\alpha - 1}{\alpha^m - 1} \, L_1 \, , \quad y_1 = \frac{\beta - 1}{\beta^n - 1} \, L_2 \, .$$

Integrating both sides of (1a) over a typical cell V, illustrated in Figure 2, and using a standard identity we obtain the relation

$$-\int_{\partial V} p_n \, ds = \iint_V f \, dxdy \, . \tag{4}$$

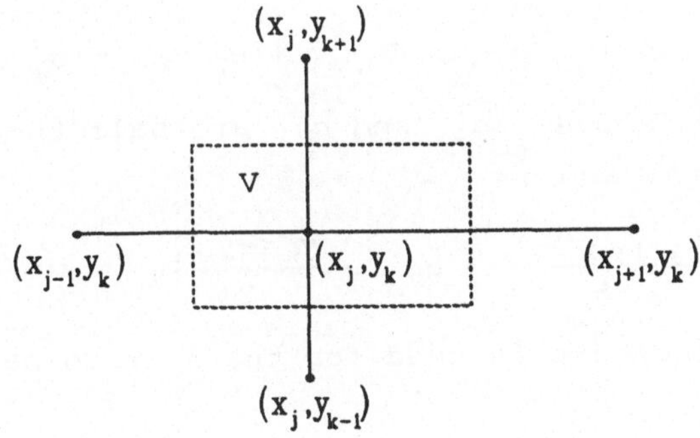

Fig. 2. A typical cell and related grid points.

Relation (4) may be discretized as follows:

$$-\left[\frac{p_{j+1,k} - p_{j,k}}{x_{j+1} - x_j} \cdot \frac{y_{k+1} - y_{k-1}}{2}\right] - \left[\frac{p_{j,k+1} - p_{j,k}}{y_{k+1} - y_k} \cdot \frac{x_{j+1} - x_{j-1}}{2}\right]$$

$$-\left[\frac{p_{j-1,k} - p_{j,k}}{x_j - x_{j-1}} \cdot \frac{y_{k+1} - y_{k-1}}{2}\right] - \left[\frac{p_{j,k-1} - p_{j,k}}{y_k - y_{k-1}} \cdot \frac{x_{j+1} - x_{j-1}}{2}\right]$$

$$\approx f_{j,k} \cdot \frac{(x_{j+1} - x_{j-1})(y_{k+1} - y_{k-1})}{4} \, .$$

This yields the algebraic equation

$$- c_W\, P_{j-1,k} - c_E\, P_{j+1,k} - c_S\, P_{j,k-1} - c_N\, P_{j,k+1} + c_Q\, P_{k,j} = b_Q$$

where

$$c_W = \frac{y_{k+1}-y_{k-1}}{2\,(x_j-x_{j-1})} = \frac{\beta^{k-1}+\beta^k}{2\alpha^{j-1}}\cdot\frac{y_1}{x_1}\ ,\quad c_E = \frac{y_{k+1}-y_{k-1}}{2\,(x_{j+1}-x_j)} = \frac{\beta^{k-1}+\beta^k}{2\alpha^j}\cdot\frac{y_1}{x_1}$$

$$c_S = \frac{x_{j+1}-x_{j-1}}{2\,(y_k-y_{k-1})} = \frac{\alpha^{j-1}+\alpha^j}{2\beta^{k-1}}\cdot\frac{x_1}{y_1}\ ,\quad c_N = \frac{x_{j+1}-x_{j-1}}{2\,(y_{k+1}-y_k)} = \frac{\alpha^{j-1}+\alpha^j}{2\beta^k}\cdot\frac{x_1}{y_1}$$

$$c_Q = c_W + c_E + c_S + c_N,\qquad b_Q = f_{j,k}\cdot\frac{(x_{j+1}-x_{j-1})\,(y_{k+1}-y_{k-1})}{4}\ .$$

If V is adjacent to $\partial\Omega$ then the relation $p_n = g$ is used for
the part of the boundary integral in (4) that involves $\partial\Omega$.
Further, since in such a cell the grid point is not centered,
some of the above coefficients are modified. For example, if
$j = 0$ and $0 < k < n$ then we have

$$- c_E\, P_{j+1,k} - c_S\, P_{j,k-1} - c_N\, P_{j,k+1} + c_Q\, P_{k,j} = b_Q$$

where c_E is as above, c_S and c_N are half the above values,
$c_Q = c_E + c_S + c_N$, and

$$b_Q = f_{0,k}\cdot\frac{x_1\,(y_{k+1}-y_k)}{4} + g_{0,k}\cdot\frac{y_{k+1}-y_{k-1}}{2}\ .$$

An analogous procedure is used for the four corner cells .

2. A singular system of equations

The finite volume method produces a system of N linear
equations in N unknowns, where $N = (m+1)(n+1)$. For
convenience we adopt the standard notation

$$\mathbf{A}\,\mathbf{x} = \mathbf{b}\ . \tag{5}$$

The matrix $\mathbf{A}$ is irreducible, symmetric, and semipositive
definite with the null space

$$\mathcal{N}(\mathbf{A}) = \{\mathbf{y}\in R^N \mid \mathbf{y} = c\mathbf{e}\ ,\quad c\in R\}$$

where $\mathbf{e} = [1,1,1,\ldots,1]^T$ .Since the diagonal entries of $\mathbf{A}$ are
positive and the off-diagonal entries are nonpositive, it
follows that $\mathbf{A}$ is a (singular) diagonally dominant M-matrix.

System (5) obviously has solutions if and only if $\mathbf{b}$
belongs to the range of $\mathbf{A}$ defined by

$$\mathscr{R}(\mathbf{A}) = \{\mathbf{z} \in R^N \mid \mathbf{A}\mathbf{y} = \mathbf{z} \text{ for some } \mathbf{y} \in R^N\}$$

Now $\mathscr{N}(\mathbf{A})$ and $\mathscr{R}(\mathbf{A})$ are orthogonal complements (because $\mathbf{A}$ is symmetric), so $\mathbf{b}$ belongs to $\mathscr{R}(\mathbf{A})$ if and only if $\mathbf{b}$ is orthogonal to $\mathscr{N}(\mathbf{A})$. Hence the consistency condition for (5) is

$$\mathbf{e}^T \mathbf{b} = \sum_{i=1}^{N} b_i = 0 \tag{6}$$

which is the discrete analog of (2).

Even though (2) is satisfied for our problems, (6) in general is not because of the integration errors inherent in the computation of $\mathbf{b}$. To obtain a consistent algebraic problem we must remove from $\mathbf{b}$ its component in $\mathscr{N}(\mathbf{A})$. This yields the vector

$$\mathbf{b}^* = \mathbf{b} - c\,\mathbf{e}, \quad c = \frac{1}{N}\,\mathbf{e}^T \mathbf{b}.$$

Since $\mathbf{b}^* \in \mathscr{R}(\mathbf{A})$, we can now address the problem of solving the singular consistent system

$$\mathbf{A}\mathbf{x} = \mathbf{b}^*. \tag{7}$$

This system has an infinity of solutions. If $\mathbf{x}_P$ is any particular solution then the set of solutions is $\mathbf{x}^* + c\mathbf{e}$ for all $c \in R$.

When a convergent iterative method is applied to (7), the solution obtained depends on the choice of the initial vector $\mathbf{x}_0$. Often one wants the solution of (7) that satisfies some specified condition (in our case, for example, this might be $p(0,0) = 1$). Extracting the desired solution from the found solution is usually a simple operation.

3. Relaxed ILU

We have solved (7) using the conjugate gradient method preconditioned by relaxed ILU. This type of incomplete factorization was investigated in [2] in the context of nonsingular problems. The code below serves to define the process. The set S determines where fill-ins are allowed. In our computations we put

$$S = \left\{(i,j) \mid a_{ij} \neq 0\right\}$$

allowing no fill-ins at all.

```
for k:= 1 to (N-1) do
   for i:= (k+1) to N do
      e := a[i,k]/a[k,k];  a[i,k] := e;
      for j:= (k+1) to N do
         if  (i,j) ∈ S then
               a[i,j] := a[i,j] - e × a[k,j]
         else
               a[i,i] := a[i,i] - e × a[k,j] .
```

The parameter ω is chosen (as best one can) to obtain a high
rate of convergence. Since the above code exploits neither
sparsity nor symmetry, it must of course be modified for
practical computation.

When it does not break down, the relaxed ILU algorithm
produces an incomplete factorization described by

$$A = LU + E = LDL^T + E \tag{8}$$

where L is lower triangular with unit entries on the main
diagonal, U is upper triangular and $D = \text{diag}(u_{11}, u_{22}, \ldots, u_{NN})$.
As implemented in the code above, the computation ends with the
strictly lower triangular part of L and the upper triangular
part of U stored in the corresponding parts of the array.

The preconditioning matrix associated with (8) is

$$M = LU = LDL^T . \tag{9}$$

An analysis of relaxed ILU yields the following information:
If $\omega < 1$ then L and U are nonsingular M-matrices and M is
symmetric positive definite. If $\omega = 1$ then L is a nonsingular
M-matrix, U is a singular M-matrix, and M is symmetric
positive semidefinite <u>with the same null space as</u> A. If $\omega > 1$
then the factorization is generally unstable and even breaks
down for certain values of ω. When it does not break down, M
is symmetric indefinite.

In the following discussion quantities dependent on ω are
so indicated by the notation. We consider the eigensolutions
of A and $M(\omega)$:

$$A u_i = \xi_i u_i, \quad M(\omega) v_i(\omega) = \lambda_i(\omega) v_i(\omega), \quad i = 1,2,\ldots,N .$$

Let the eigensolutions of A be ordered by

$$0 = \xi_1 < \xi_2 \leq \cdots \leq \xi_N$$

and those of $\mathbf{M}(\omega)$ by

$$\lambda_1(\omega) \le \lambda_2(\omega) \le \ldots \le \lambda_N(\omega) .$$

We have $\lambda_1(\omega) > 0$ if $\omega < 1$ and $\lambda_1(\omega) = 0$ if $\omega = 1$. Each set of eigenvectors may be assumed to be orthonormal, and we have the important relation

$$\mathbf{u}_1 = \mathbf{v}_1(1) = \frac{1}{\sqrt{N}}\, \mathbf{e} .$$

The rate of convergence of the preconditioned conjugate gradient method depends on the eigenvalues of $\mathbf{M}^{-1}(\omega)\mathbf{A}$. Since $\mathbf{M}(1)$ is singular, it is relevant to examine the behavior of $\mathbf{M}^{-1}(\omega)\mathbf{A}$ as $\omega \to 1$. To this end we apply the spectral decompositions

$$\mathbf{M}^{-1}(\omega) = \sum_{i=1}^{N} \lambda_i^{-1}(\omega)\,\mathbf{v}_i(\omega)\,\mathbf{v}_i^{T}(\omega), \qquad \mathbf{A} = \sum_{j=2}^{N} \xi_j\, \mathbf{u}_j \mathbf{u}_j^{T}$$

to obtain, after some rearranging,

$$\mathbf{M}^{-1}(\omega)\,\mathbf{A} = \sum_{j=2}^{N} \left[\phi_j(\omega)\cdot\xi_j\cdot\mathbf{v}_1(\omega)\,\mathbf{u}_j^{T} \right] + \mathbf{C}(\omega) \tag{10}$$

where

$$\phi_j(\omega) = \frac{\mathbf{v}_1^{T}(\omega)\,\mathbf{u}_j}{\lambda_1(\omega)} , \quad 2 \le j \le n \tag{11}$$

and $\mathbf{C}(\omega)$ is regular in a neighborhood of $\omega = 1$.

Since $\lambda_1(1) = 0$ and $\mathbf{v}_1^{T}(1)\mathbf{u}_j = 0$, $2 \le j \le n$, the behavior of $\phi_j(\omega)$ as $\omega \to 1$ needs further attention. Applying first-order perturbation theory (see [5], for example; we omit the details) one finds that $\phi_j(\omega)$ converges and hence also $\mathbf{M}^{-1}(\omega)\mathbf{A}$.

For example, if $L_1 = L_2 = m = n = 1$ then for $\omega \ne 3$,

$$\mathbf{M}^{-1}(\omega)\,\mathbf{A} = \begin{bmatrix} 1 & -\dfrac{1}{2} & -\dfrac{1}{2} & 0 \\[2ex] 0 & \dfrac{2}{3-\omega} & -\dfrac{2}{3-\omega} & 0 \\[2ex] 0 & -\dfrac{2}{3-\omega} & \dfrac{2}{3-\omega} & 0 \\[2ex] 0 & -\dfrac{1}{2} & -\dfrac{1}{2} & 1 \end{bmatrix}$$

with eigenvalues $0, 1, 1$ and $4/(3-\omega)$.

The analysis of singular preconditioning can be nicely formulated in terms of generalized matrix inverses. For this as well as much other relevant material see [3] and [4].

We turn now to the implementation of the preconditioned conjugate gradient method as applied to (7). An algorithm may be found, for example, on p. 78 of [1]. The preconditioning matrix $\mathbf{M}$ arises in the step

$$\mathbf{M}\mathbf{h}_k = \mathbf{r}_k \quad (\text{solve for } \mathbf{h}_k)$$

where $\mathbf{r}_k = \mathbf{b}^* - \mathbf{A}\mathbf{x}_k$. With $\mathbf{M} = \mathbf{L}\mathbf{D}\mathbf{L}^T$ (see (9)) the system can be decomposed as follows:

$$\text{a)} \quad \mathbf{L}\mathbf{y}_k = \mathbf{r}_k \qquad \text{b)} \quad \mathbf{D}\mathbf{z}_k = \mathbf{y}_k \qquad \text{c)} \quad \mathbf{L}^T\mathbf{h}_k = \mathbf{z}_k .$$

When $\omega = 1$ then b) is singular with $d_{ii} > 0$, $i = 1, 2, \ldots, N-1$ and $d_{NN} = 0$. It is easy to show that the consistency of (7) leads to the consistency of b); i.e. the last component of $\mathbf{y}_k$ is zero. Hence $(\mathbf{z}_k)_i = (\mathbf{y}_k)_i/d_{ii}$, $i = 1, 2, \ldots, N-1$, and $(\mathbf{z}_k)_N$ can be chosen arbitrarily.

4. Numerical results

For any matrix $\mathbf{B}$ with real nonnegative eigenvalues we define $\kappa(\mathbf{B})$ to be the largest eigenvalue divided by the smallest nonzero eigenvalue. In Figure 4 we show $\kappa(\mathbf{A})$, $\kappa(\mathbf{A}_{DS})$ and $\kappa(\mathbf{M}^{-1}\mathbf{A})$ for a series of grids. Here $\mathbf{A}$ is the coefficient matrix in (7), $\mathbf{A}_{DS}$ is the matrix obtained by a symmetric diagonal scaling of $\mathbf{A}$, i.e. $(\mathbf{A}_{DS})_{ij} = a_{ij}/(a_{ii} a_{jj})^{1/2}$ and $\mathbf{M}$ is given by (9) with $\omega = 0.5$ The grid parameters are $L_1 = L_2 = 1$, $m = n = 15$, and we have put $\beta = \alpha$. All matrices are of order 256.

Figure 3 shows that $\kappa(\mathbf{A})$ increases rapidly as α increases; i.e., $\mathbf{A}$ becomes poorly conditioned as the grid is deformed. Diagonal scaling helps somewhat, but the tendency is the same. In contrast, $\kappa(\mathbf{M}^{-1}\mathbf{A})$ is remarkably unaffected by α and remains small in the range of α considered.

In Figure 4 we show the eigenvalues of $\mathbf{M}^{-1}\mathbf{A}$ for $\alpha = 1, 2, 3$ and 5. We see that as α increases, there is an increasing clustering of eigenvalues about unity.

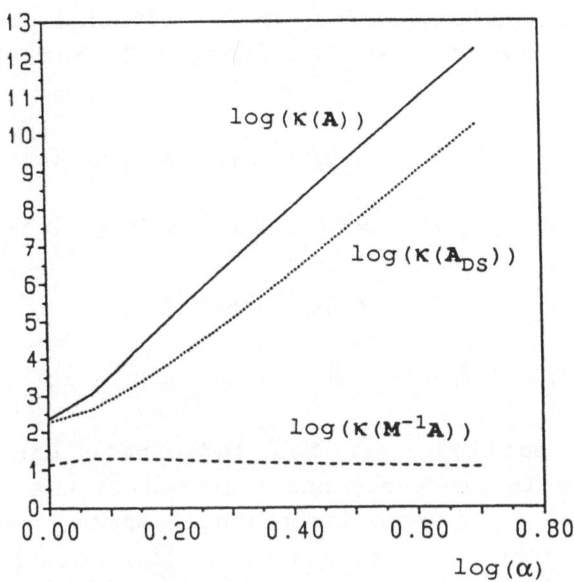

Fig. 3. Condition numbers of $\mathbf{A}$, $\mathbf{A}_{DS}$ and $\mathbf{M}^{-1}\mathbf{A}$.

Fig. 4. Spectrum of $\mathbf{M}^{-1}\mathbf{A}$ for $\alpha = 1,2,3$ and 5.

Because of the spectral properties of $\mathbf{M}^{-1}\mathbf{A}$ displayed in Figures 3 and 4, one expects relaxed ILU to be an effective preconditioner for the conjugate gradient method applied to (7). We have solved (7) using the conjugate gradient method with and without relaxed ILU preconditioning. We chose $\mathbf{b}^{*}=\mathbf{A}\mathbf{x}_P$, with $(\mathbf{x}_p)_i = \sin(i)$. The initial vector was $\mathbf{x}_0 = \mathbf{0}$.

Four cases were considered, these differing with respect to the use or nonuse of preconditioning (PC) and the choice of stopping test:

No PC; $$\| \mathbf{x}_k - \mathbf{x} \|_A / \| \mathbf{x}_0 - \mathbf{x} \|_A \leq 10^{-6} \qquad (12a)$$

" "; $$\| \mathbf{x}_k - \mathbf{x} \|_\infty / \| \mathbf{x}_0 - \mathbf{x} \|_\infty \leq 10^{-6} \qquad (12b)$$

Relaxed ILU PC; $$\| \mathbf{x}_k - \mathbf{x} \|_A / \| \mathbf{x}_0 - \mathbf{x} \|_A \leq 10^{-6} \qquad (12c)$$

" " "; $$\| \mathbf{x}_k - \mathbf{x} \|_\infty / \| \mathbf{x}_0 - \mathbf{x} \|_\infty \leq 10^{-6}. \qquad (12d)$$

As mentioned earlier, (7) has infinitely many solutions. However, there is precisely one solution in the range $\mathcal{R}(\mathbf{A})$, and we obtained it by removing the null space component of $\mathbf{x}_p$. This is the vector $\mathbf{x}$ referred to in (12). Similarly, we have removed the null space component of each $\mathbf{x}_k$. These operations are superfluous in the cases of (12a) and (12c) since null space components vanish in the computation of the $\mathbf{A}$-norm (actually seminorm here). They are needed for the infinity norm, however.

Table 1 gives the number of iterations for each of the four cases when α varies between 1 and 5. We see that relaxed ILU does indeed give a dramatic acceleration of convergence.

Table 1. Number of iterations

α	(12a)	(12b)	(12c)	(12d)
1.000	53	56	14	19
1.175	102	118	15	17
1.380	216	258	13	16
1.621	444	533	12	14
1.904	845	1130	10	13
2.236	1602	2189	9	13
2.627	2456	4062	7	12
3.085	2924	7439	5	12
3.624	3406	-	3	11
4.257	4721	-	3	11
5.000	5165	-	2	11

Initially we investigated only (12a) and (12c). However, it was observed that as α increased, the accuracy of the accepted solution vector $\mathbf{x}_k$ deteriorated with respect to the infinity norm (possibly because of rounding errors). Therefore we added (12b) and (12d). Our computations were performed on an AMDAHL VP1100 vector processor in double precision (about 16 decimal digits).

References

[1] AXELSSON, O., BARKER, V.A.: "Finite Element Solution of Boundary Value Problems", Academic Press, Orlando, Florida, 1984.

[2] AXELSSON, O., LINDSKOG, G.: "On the eigenvalue distribution of a class of preconditioning methods", Numer. Math., **48** (1986) pp. 479-498.

[3] NOTAY, Y.: "Incomplete factorization of singular linear systems", BIT, **29** (1989) pp. 682-702.

[4] NOTAY, Y. : "Solving positive (semi)definite linear systems by preconditioned iterative methods", in Preconditioned Conjugate Gradient Methods, O. Axelsson and L. Kolotilina, eds., LNiM No.1457 pp. 105-125, Springer-Verlag, 1990.

[5] WILKINSON, J.H.: "The Algebraic Eigenvalue Problem", Clarendon Press, Oxford, 1965.

Locally Refined Solution of Unsymmetric and Nonlinear Problems[*]

Peter Bastian[#]

Interdisziplinäres Zentrum für Wissenschaftliches Rechnen,
Universität Heidelberg, Im Neuenheimer Feld 368, D-6900 Heidelberg

Abstract

Abstract   We describe a multigrid method with optimal computational work per cycle on locally refined grids. The method can be interpreted as a multiplicative variant of the BPX preconditioner but it is motivated from the viewpoint of the classical multigrid method. This has several advantages: In the case of quasi-uniform refinement the method is equivalent to the classical multigrid method. All well known smoothing algorithms can be used, including incomplete decompositions. In the nonlinear case the nonlinear multigrid method can be directly transferred to locally refined grids. Since no outer CG iteration is needed the method can also be applied directly to unsymmetric problems. Results will be presented for scalar, linear and nonlinear convection-diffusion equations.

Keywords   multigrid method, unstructured, locally refined grids, nonlinear p.d.e, unsymmetric linear systems

1   Introduction

The multigrid method has been designed as an optimal method for scalar elliptic PDE, see [10],[17] for an introduction to multigrid methods. Meanwhile it has been applied with great success also to systems of nonlinear PDE's like the incompressible, stationary Navier-Stokes equations [4],[13],[18]. This has been achieved by combining the multigrid method with robust smoothing iterations in the singularly perturbed cases of dominating convection and anisotropic diffusion.

Robust smoothers are exact (or very fast) single grid solvers in the limit case. Good candidates are the block line Gauß-Seidel method or the incomplete decomposition method ([14],[11],[19]). Especially the latter method is very interesting since it is able to solve upper or lower triangular and tridiagonal linear systems exactly. Linear systems with such a pattern often arise in practice when an upwind discretization is used for dominating convective terms (triangular systems) or anisotropic diffusion is present (tridiagonal systems) *and* a structured grid with appropriate node ordering is used. It is important to notice that the structure of the matrix depends on both, the structure of the grid and the node ordering, which makes this approach difficult for general unstructured meshes.

---

*. This work has been supported by the Deutsche Forschungsgemeinschaft (DFG).

#. email address: bastian@iwr1.iwr.uni-heidelberg.de

The methods mentioned so far have been used mainly on structured, globally refined grids. Another branch of multigrid (often called multilevel in this context) methods evolved for the solution of elliptic PDE's on locally refined, unstructured meshes. There Yserentant ([21]) and Bank, Dupont, Yserentant ([1],[2]) developed almost optimal methods for two space dimensions and recently Bramble, Pasciak and Xu ([7]) developed an optimal method also for higher dimensions (see also the proofs in [5],[6],[15]).

In the terminology of Xu ([20]) the Hierarchical Basis preconditioner (HB) and the BPX preconditioner are additive methods, which means that the global defect is restricted to all coarse grid spaces (imagine $v_1=0$ in classical multigrid), then smoothing is performed on all coarser levels in parallel, where the smoother is a simple Richardson (with special weighting factors) or Jacobi iteration ([22]) and finally all corrections are summed up. This iteration alone is not convergent, but it can be used as a preconditioner for the conjugate gradient method. In the HB method each node is visited exactly once at the level it has been introduced during the refinement process, whereas in the BPX method a node is visited on each level as long as it is in the region of refinement.

The Hierarchical Basis Multigrid Method (HB/MG) from [1] is the multiplicative version of the HB method and is convergent without the outer CG iteration. In [2] it is also used with more complicated smoothers like symmetric Gauß-Seidel. All three methods mentioned work poorly or not at all for unsymmetric linear systems arising from an upwind discretization of the convection-diffusion equation. HB and BPX are not directly applicable because of the outer CG iteration and the simple jacobi type smoother will not be efficient even if the CG method is replaced by an unsymmetric CG variant. HB/MG performs poorly because "too few" grid points are smoothed on each level (see results section).

The method described in the sequel could be described as a multiplicative multigrid algorithm with "local" smoothing, which means that smoothing is restricted to refined regions only. Such a method has been considered theoretically in [8], and it could also be interpreted as the multiplicative variant of the BPX algorithm, i. e. it is to BPX what HB/MG is to the HB method. In combination with a robust smoother it is able to solve unsymmetric systems of the type mentioned above very efficiently. In addition the method is motivated from the viewpoint of the classical multigrid method, which makes it possible to formulate an analogue to the nonlinear multigrid method and also gives some hints for the implementation.

The next section will describe the linear and the nonlinear version of the multigrid method with local smoothing and the third section shows some numerical results.

2   Multigrid with local smoothing

2.1   Discretization and grid hierarchy

We consider the following general, scalar PDE

$$\mathrm{div}\,(\vec{r}(x, y, u)u - \varepsilon(x, y, u, u_x, u_y)\nabla u) = f(x, y) \tag{1}$$

in a domain $\Omega \subseteq \Re^2$, with dirichlet boundary conditions $u(x, y) = g(x, y)$ on $\partial\Omega$. Note that eq. (1) may be nonlinear. In section 3 we will show numerical results for two different equations in this class.

The PDE is discretized on unstructured meshes consisting of triangles and convex quadrangles using a finite volume method following [12] and [16]. The first order terms are discretized by a simple upwind technique, where the basis functions (for first order terms only) are assumed to be piecewise constant over each box and the line integral over the box boundary uses the left or right value depending on the local value of $\hat{r}$ (flow direction).

The mesh hierarchy is built as follows. We start with a coarse triangulation $\mathcal{T}_0$ covering the domain Ω and the usual rule that two elements can have at most a vertex or a side in common. Successively finer triangulations $\mathcal{T}_1$, $\mathcal{T}_2$, ..., $\mathcal{T}_j$ are obtained by either copying old elements or refining selected elements according to some refinement criterion. The possible refinement rules implemented are shown in Figure 1. The irregular refinement rules are necessary to get conforming triangulations in case of local refinement and the resulting irregular elements must not be further refined (*nested* triangulation) in order to bound the smallest angle. Each element e is assigned a unique level. Since an element may be in more than one $\mathcal{T}_k$ it is assigned the smallest k where $e \in \mathcal{T}_k$ holds. Note that each $\mathcal{T}_k$ covers the whole domain Ω, i.e. in case of local mesh refinement $\mathcal{T}_k$ can contain elements with level less than or equal to k.

The set $\mathcal{V}_k$ consists of all vertices of triangulation $\mathcal{T}_k$. A vertex v may be in more than one $\mathcal{V}_k$. The index of the first $\mathcal{V}_k$ containing v is called the level of v.

Figure 1   Refinement rules implemented in ug 2.0.

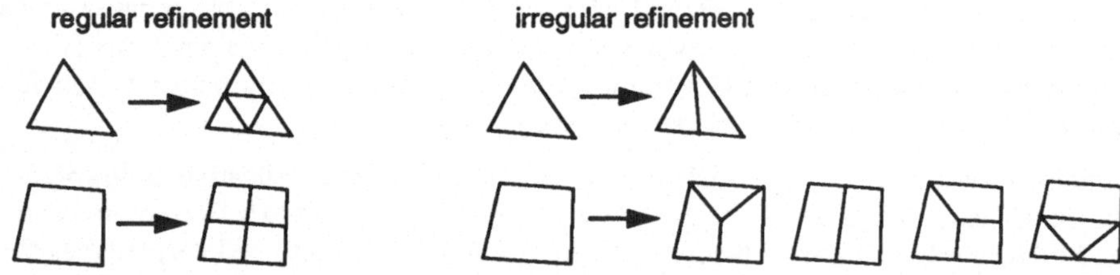

Figure 1   Refinement rules implemented in ug 2.0.

2.2   Linear algorithm

Now imagine the standard multigrid algorithm on the sequence of grids $\mathcal{T}_0$, .., $\mathcal{T}_j$, which upon entry on level k would smooth all vertices $v \in \mathcal{V}_k$, compute the defect, transfer the defect to $\mathcal{T}_{k-1}$ using the canonical restriction, call the algorithm recursively ($k>0$), prolongate the correction from $\mathcal{T}_{k-1}$ and finally corrects the solution. The problem with that algorithm is that its work count is not optimal since a geometric growth in the number of vertices

$$\# \mathcal{V}_k \geq C \cdot \# \mathcal{V}_{k-1} \tag{2}$$

with $C>1$ for all k does not hold for arbitrary local refinement. In addition also the memory requirements of that algorithm are not proportional to $N = \# \mathcal{V}_j$ any more.

The solution of the HB and HB/MG methods to that problem was to smooth only the level k vertices on level k. However, it turned out that more vertices have to be smoothed in order to get a rate of convergence independent of refinement depth. Let $\mathcal{W}_k$ be the set of level k vertices:

$$\mathcal{W}_k = \mathcal{V}_k \backslash \bigcup_{i=0}^{k-1} \mathcal{V}_i \tag{3}$$

and let the relation R_k on $\mathcal{V}_k$ be given by

$$(v, v') \in R_k \Leftrightarrow v, v' \text{ are corners of an element } e \in \mathcal{T}_k, \qquad (4)$$

then all $v \in S_k$ are smoothed on level k in our algorithm with S_k given by

$$S_0 = \mathcal{V}_0$$
$$S_k = \mathcal{W}_k \cup \{ v \in \mathcal{V}_k \mid (\exists (v' \in \mathcal{W}_k)) \, ((v, v') \in R_k) \} \qquad (k > 0). \qquad (5)$$

The work count of this algorithm is still optimal since on each level the work for smoothing is proportional to $n_k = \# \mathcal{W}_k$ and $n_0 + \ldots + n_j = N$.

For the full description of our multigrid algorithm we also need the subset $\mathcal{D}_k$ of $\mathcal{V}_k$ given by

$$\mathcal{D}_k = S_k \cup v \in \mathcal{V}_k \mid (\exists (v' \in S_k)) \, ((v, v') \in R) \qquad . \qquad (6)$$

The discretization of eq. (1) on each $\mathcal{T}_k$ generates linear systems of equations denoted by

$$A_k x_k = b_k \qquad . \qquad (7)$$

We further assume that an ordering is defined on the vertices, given by a function index: $\mathcal{V}_j \to \{1, 2, \ldots, N\}$ with index(v)<index(v') if level(v)<level(v'). All vectors used in this paper are assumed to be ordered according to this mapping. This means that the i'th components of x_k and x_{k+1} correspond to the same vertex v_i and are denoted by $x_{k,i}$ and $x_{k+1,i}$ respectively. We are now able to formulate the linear algorithm:

Algorithm 1   Linear multigrid algorithm with local smoothing. Assume x_j has been computed.

```
        void lmgc (int j, int k)
        {
            if (k==j)
            { /* compute initial defect */
```

(1.1) $\forall l$, set $d_{l,i} = b_{l,i} - \sum_{a_{l,im} \neq 0} a_{l,im} x_{j,m}$ if vertex $v_i \in \mathcal{D}_l \setminus \mathcal{D}_{l+1}$ $(\mathcal{D}_{j+1} = \varnothing)$ else $d_{l,i} = 0$;

(1.2) $\forall l$, set $y_l = 0$.

```
            }
```

(1.3)   if (k==0) { $y_0 = A_0^{-1} d_0$; }

     else {

(1.4)   Smooth $A_k y_k = d_k$ v_1 times at vertices $v \in S_k$ only;

(1.5)   Compute $z_{k,i} = d_{k,i} - \sum_{a_{k,im} \neq 0} a_{k,im} y_{k,m}$ if vertex $v_i \in \mathcal{D}_k$, else $z_{k,i} = 0$;

(1.6)   restrict $d_{k-1} = d_{k-1} + r z_k$. r is the canonical restriction;

(1.7)   lmgc(j,k-1);

(1.8)   Interpolate $z_k = p \, y_{k-1}$ at all vertices $v \in \mathcal{D}_k$;

(1.9)   Add $y_k = y_k + z_k$ at all vertices $v \in \mathcal{D}_k$;

(1.10)   Smooth $A_k y_k = d_k$ v_2 times at vertices $v \in S_k$ only;

     }

(1.11)   Final correction: $x_{j,i} = x_{j,i} + y_{k,i}$ if $v_i \in S_k \setminus S_{k+1}$ (where $S_{j+1} = \varnothing$);

   }

The local smoothing at vertices $v \in S_k$ on level k has the immediate consequence that the defect changes only at the vertices $v \in \mathcal{D}_k$. Outside $\mathcal{D}_k$ the defect is equal to the defect at the beginning of the whole cycle. Step (1.1) initializes the defect values where they are needed for the first time in the course of the algorithm. Step (1.3) is the exact solution of the coarse grid problem. Steps (1.4) and (1.10) are the usual pre- and post smoothing steps with an iterative method like Gauß-Seidel or ILU but only values at the vertices contained in S_k are allowed to change. As a consequence step (1.5) computes the new defect only at the locations where it changed with respect to the beginning of the cycle. Step (1.7) allows only one recursive call (V-cycle) since more than one call would give a non optimal work count. Steps (1.8)/(1.9) update the values of y_k only where they are needed in the following post-smoothing step. Step (1.11) updates the values of x_j, the fine grid solution, at those vertices where the final correction has been computed (since the smoothers on the finer levels will not change these values).

In order to carry out algorithm 1, *only* the vertices of $\mathcal{M}_k \subseteq \mathcal{T}_k$ given by

$$\mathcal{M}_k = \mathcal{D}_k \cup \{v \in \mathcal{V}_k | \ (\exists (v' \in \mathcal{D}_k)) \ ((v,v') \in R_k) \} \tag{8}$$

and the elements of $\mathcal{E}_k \subseteq \mathcal{T}_k$ given by

$$\mathcal{E}_k = \{e \in \mathcal{T}_k | e \text{ has a corner } v \text{ in } \mathcal{D}_k\} \tag{9}$$

have to be stored on level k. From $\mathcal{E}_k$ all matrix elements needed in algorithm 1 can be assembled. It can be shown that

$$\# \mathcal{M}_k \le C(\mathcal{T}_0) \, n_k \tag{10}$$

holds and therefore $\sum_{k=0}^{j} \# \mathcal{M}_k \le C(\mathcal{T}_0) \, N$, which means that also memory requirements are optimal.

2.3 Nonlinear algorithm

From the linear algorithm we can develop the nonlinear algorithm. The nonlinear multigrid algorithm is of particular interest since it makes no explicit use of the Jacobian. The discretization of eq. (1) in the nonlinear case on the triangulations $\mathcal{T}_k$ yields nonlinear discrete systems of the form

$$A_k(x_k)x_k = b_k \quad . \tag{11}$$

The basic idea of the nonlinear multigrid method (chap. 9 in [10]) is to solve the coarse grid problem

$$A_{k-1}(y_{k-1})y_{k-1} = A_{k-1}(x^*_{k-1})x^*_{k-1} + s_{k-1}rd_k \tag{12}$$

with d_k the defect on the fine grid, and x^*_{k-1} the numerical solution of (11) obtained from the nested iteration process. The scalar s_{k-1} is chosen small enough that (12) is invertible and is different on each level depending on $\|rd_k\|_*$ (euclidian norm). The correction v_{k-1} is computed from y_{k-1} as $v_{k-1} = s_{k-1}^{-1}(y_{k-1} - x^*_{k-1})$.

The nonlinear multigrid method needs a nonlinear single grid method as smoother. For that purpose we take a simple damped fixed point iteration:

$$x_k^{n+1} = x_k^n + \omega A_k^{-1}(x_k^n)(b_k - A_k(x_k^n)x_k^n) \tag{13}$$

where the inversion of $A_k(x_k^n)$ is replaced by ν steps of a linear single grid method (GS, ILU). In section 3 the fixed point iteration is also used as a stand alone solution method with the inversion of $A_k(x_k^n)$ replaced by one or several steps of algorithm 1.

We now formulate the nonlinear version of algorithm 1:

Algorithm 2   Nonlinear multigrid with local smoothing.

```
        void nlmgc (int j, int k, double q)
        {
            double s;
            if (k==j)
            {
```

(2.1)   Set $d_{j,i} = b_{j,i}$ if $v_i \in \mathcal{D}_j$ and $y_{j,i} = x_{j,i}$ if $v_i \in \mathcal{M}_j$ or 0 else;

(2.2) $\forall\,(l<j)$, set $d_{l,i} = b_{l,i} - \displaystyle\sum_{a_{l,im} \neq 0} a_{l,im}(x_j)x_{j,m}$ if $v_i \in \mathcal{D}_l \backslash \mathcal{D}_{l+1}$;

           }

(2.3)   if ($k==0$) { Solve $A_0(y_0)y_0 = d_0$; }

           else {

(2.4)   Nonlinear smoothing of $A_k(y_k)y_k = d_k$ at vertices $v \in \mathcal{S}_k$ only;

(2.5)   Compute $z_{k,i} = d_{k,i} - \displaystyle\sum_{a_{k,im} \neq 0} a_{k,im}(y_k)y_{k,m}$ if vertex $v_i \in \mathcal{D}_k$, or $z_{k,i} = 0$ else;

(2.6)   Compute $s = \sigma/\|z_k\|$;

(2.7)   Prepare coarse grid problem:

$$d_{k-1,i} = \sum_{a_{k-1,im} \neq 0} a_{k-1,im}(x_{k-1}^*)x_{k-1,m}^* + \begin{cases} s(rz_k)_i & v_i \in \mathcal{D}_k \\ qsd_{k-1,i} & v_i \in \mathcal{D}_{k-1} \backslash \mathcal{D}_k \end{cases}$$

(2.8)   Set $y_{k-1,i} = x_{k-1,i}^*$ for $v_i \in \mathcal{M}_{k-1}$

(2.9)   nlmgc($j,k-1,qs$);

(2.10)   Compute $y_{k-1,i} = y_{k-1,i} - x_{k-1,i}$ for all $v_i \in \mathcal{M}_{k-1}$

(2.11)   Interpolate $z_k = p\, y_{k-1}$ at all vertices $v \in \mathcal{M}_k$;

(2.12)   Add $y_k = y_k + s^{-1}z_k$ at all vertices $v \in \mathcal{M}_k$;

(2.13)   Nonlinear smoothing of $A_k(y_k)y_k = d_k$ at vertices $v \in \mathcal{S}_k$ only;

           }

(2.14)   Final correction: $x_{j,i} = x_{j,i} + q^{-1}y_{k,i}$ if $v_i \in \mathcal{S}_k \backslash \mathcal{S}_{k+1}$ (where $\mathcal{S}_{j+1} = \varnothing$);

           }

Step (2.6) is the computation of the scalar s_k, σ is a small constant. Step (2.7) prepares the coarse grid problem and needs the product of all previous values of s stored in q in order to scale the defect correctly. Steps (2.8) and (2.10)-(2.12) involve the sets $\mathcal{M}_k$ in order to compute $A_k(y_k)$ correctly.

Algorithm 2 has one conceptual problem. The grid refinement process may change any of the triangulations $\mathcal{T}_1, ..., \mathcal{T}_j$ in order to ensure the nestedness of the triangulation. In that case the solution x_k^* is not available at these new vertices. There are two possibilities to remedy the situation: First one could start a new nested iteration from the lowest level where the grid changed, but this will be very inefficient in most cases. The other possibility is to replace x_k^* from the nested iteration by the restricted current fine grid solution x_j which resembles Brandt's FAS multigrid algorithm [9]. This version will be used in section 3.2.

2.4 Implementation remarks

The algorithms explained above have been implemented in the flexible unstructured grid code *ug 2.0* [3], which is available freely from the author. The sets $\mathcal{W}_k$, $\mathcal{S}_k$, $\mathcal{D}_k$, $\mathcal{M}_k$ and $\mathcal{E}_k$ can be computed easily during the refinement process. In order to store $\mathcal{E}_k$, a dummy refinement rule that simply copies an element has been added to those depicted in Figure 1.

3 Numerical results

This section presents numerical results for a linear and a nonlinear convection-diffusion example.

3.1 Linear convection-diffusion

The equation

$$c\,(\cos\alpha,\,\sin\alpha)\,\nabla u - \Delta u = 0 \tag{14}$$

in $\Omega = (0,1)^2$ is used as test example for the linear algorithm. The boundary conditions are of dirichlet type:

$$u\,(x,0) = \left\{ \begin{array}{ll} 1 & |x-0.5| < 0.1 \\ 0 & \text{else} \end{array} \right. \qquad u\,(x,1) = u\,(1,y) = 1 \qquad u\,(0,y) = 0. \tag{15}$$

$\alpha = 3\pi/8$ is used in all examples. The solution develops a boundary layer at the north and east boundaries and two interior layers due to the jumps in the boundary conditions. The discretization used is a finite volume method with first order upwinding for the convective terms as described earlier. The coarsest grid $\mathcal{T}_0$ in all computations was the standard triangulation with h=1/2 obtained by dividing the unit square into four equal quadrangles and then subdividing each quadrangle by connecting the lower left and the upper right corner. In local refinement calculations $\mathcal{T}_1$ and $\mathcal{T}_2$ are obtained by global refinement and the $\mathcal{T}_k$, k>2, are determined by refining all elements e where $\| \nabla u \|_e \geq 3.0$.

Table 1 shows average convergence rates of algorithm 1 with a Gauß-Seidel smoother compared to the HB/MG method simulated with algorithm 1 by restricting smoothing on level k only to vertices in $\mathcal{W}_k$. The vertices were ordered lexicographically (left to right bottom to top) in the first part of the table and exactly reverse in the second part of the table. With lexicographic ordering Gauß-Seidel is a very good solver for the single grid problem, since the matrices are almost triangular. Algorithm 1 one combined with such a smoother is very efficient just as in the structured grid case and the convergence rates even get better with the convection dominating the equation. With the ordering reversed the convergence rates are much worse, as expected. In the HB/MG method the ordering of the unknowns makes no difference. The reason for that is that vertices in $\mathcal{V}_k \backslash \mathcal{W}_k$ are only changed through the coarse grid correction process which does not work for this unsymmetric problem. It should be noted however, that in general a good solver for the single grid problem is not so easy to find as in this example.

Table 1   Linear convection-diffusion calculations with local grid refinement. Multigrid parameters where $\nu_1=1$, $\nu_2=1$, numbers given are average convergence rates defined by $\bar{\rho} = \sqrt[n]{\|r_n\|/\|r_0\|}$, where r_i is the residual after the i'th iteration and n is the number of iterations needed to get a residual reduction of 10^{-6} or 60 iterations whatever criterion is reached first.

		Gauß-Seidel smoother on $\mathcal{S}_k$			Gauß-Seidel smoother on $\mathcal{W}_k$		
	level	$c = 10^2$	$c = 10^4$	$c = 10^6$	$c = 10^2$	$c = 10^4$	$c = 10^6$
lexicographic ordering	1	$8.47 \cdot 10^{-3}$	$1.05 \cdot 10^{-3}$	$1.00 \cdot 10^{-3}$	$4.58 \cdot 10^{-1}$	$4.70 \cdot 10^{-1}$	$4.71 \cdot 10^{-1}$
	2	$3.10 \cdot 10^{-2}$	$4.92 \cdot 10^{-3}$	$4.88 \cdot 10^{-3}$	$6.87 \cdot 10^{-1}$	$7.13 \cdot 10^{-1}$	$7.13 \cdot 10^{-1}$
	3	$5.10 \cdot 10^{-2}$	$8.58 \cdot 10^{-2}$	$9.91 \cdot 10^{-2}$	$6.73 \cdot 10^{-1}$	$6.85 \cdot 10^{-1}$	$6.85 \cdot 10^{-1}$
	4	$8.10 \cdot 10^{-2}$	$1.53 \cdot 10^{-2}$	$1.43 \cdot 10^{-2}$	$6.33 \cdot 10^{-1}$	$7.22 \cdot 10^{-1}$	$7.22 \cdot 10^{-1}$
	5	$1.36 \cdot 10^{-1}$	$2.01 \cdot 10^{-2}$	$1.77 \cdot 10^{-2}$	$7.77 \cdot 10^{-1}$	$8.47 \cdot 10^{-1}$	$8.48 \cdot 10^{-1}$
	6	$1.82 \cdot 10^{-1}$	$2.90 \cdot 10^{-2}$	$2.23 \cdot 10^{-2}$	$8.31 \cdot 10^{-1}$	$8.94 \cdot 10^{-1}$	$8.94 \cdot 10^{-1}$
	7	$1.96 \cdot 10^{-1}$	$5.38 \cdot 10^{-2}$	$3.60 \cdot 10^{-2}$	$8.44 \cdot 10^{-1}$	$9.03 \cdot 10^{-1}$	$9.03 \cdot 10^{-1}$
reversed lexicographic ordering	1	$1.63 \cdot 10^{-1}$	$1.79 \cdot 10^{-1}$	$1.79 \cdot 10^{-1}$	$4.70 \cdot 10^{-1}$	$4.80 \cdot 10^{-1}$	$4.80 \cdot 10^{-1}$
	2	$3.09 \cdot 10^{-1}$	$3.44 \cdot 10^{-1}$	$3.43 \cdot 10^{-1}$	$7.10 \cdot 10^{-1}$	$7.37 \cdot 10^{-1}$	$7.38 \cdot 10^{-1}$
	3	$2.47 \cdot 10^{-1}$	$2.71 \cdot 10^{-1}$	$2.70 \cdot 10^{-1}$	$6.85 \cdot 10^{-1}$	$6.86 \cdot 10^{-1}$	$6.87 \cdot 10^{-1}$
	4	$2.28 \cdot 10^{-1}$	$3.64 \cdot 10^{-1}$	$3.64 \cdot 10^{-1}$	$6.88 \cdot 10^{-1}$	$7.44 \cdot 10^{-1}$	$7.44 \cdot 10^{-1}$
	5	$2.60 \cdot 10^{-1}$	$4.81 \cdot 10^{-1}$	$4.91 \cdot 10^{-1}$	$7.94 \cdot 10^{-1}$	$8.60 \cdot 10^{-1}$	$8.60 \cdot 10^{-1}$
	6	$2.91 \cdot 10^{-1}$	$6.53 \cdot 10^{-1}$	$6.43 \cdot 10^{-1}$	$8.40 \cdot 10^{-1}$	$8.98 \cdot 10^{-1}$	$8.99 \cdot 10^{-1}$
	7	$2.83 \cdot 10^{-1}$	$7.48 \cdot 10^{-1}$	$7.42 \cdot 10^{-1}$	$8.53 \cdot 10^{-1}$	$9.06 \cdot 10^{-1}$	$9.06 \cdot 10^{-1}$

3.2   Nonlinear convection-diffusion

A small modification of eq. (14) yields the nonlinear equation

$$cu(\cos\alpha, \sin\alpha)\nabla u - \Delta u = 0 \tag{16}$$

in $\Omega = (0,1)^2$ with the same boundary conditions as in the linear example. This equation was chosen because the nonlinearity resembles that of the Navier-Stokes equations.

Table 2 compares algorithm 2 and a stand alone fixed point iteration (eq. (13)) with linear equations solved by algorithm 1 for local and global grid refinement and various values of c. The upper and lower part of the table show that convergence rates are comparable for local and global refinement. Damping with a factor of $\omega=0.7$ was necessary to achieve convergence of the nonlinear multigrid method (smoother and coarse grid correction) when c was greater or equal to 10^3 and also the stand alone fixed point iteration had to be damped ($\omega=0.7$) if c was greater or equal 10^2. In general the outer fixed point iteration was much more insensitive to parameters c and ω and is also more efficient in terms of computer time since algorithm 2 must reassemble (parts of) the stiffness matrix at least three times on each level. On the other hand the inner solves in the stand alone fixed point iteration should be quite exact for larger values of c. We used a stopping criterion of 10^{-4} for the inner solves which required 3 to 4 iterations with algorithm 1 (convergence rate was always below 0.1).

Nonlinear convection-diffusion calculation. Multigrid parameters were $v_1=1$, $v_2=1$ and vertices were ordered lexicgraphically for both methods. Nonlinear multigrid used an ILU method as solver for the fixed point iteration used as smoother. The value of σ was $5 \cdot 10^{-4}$. The stand alone fixed point iteration used algorithm 1 with ILU smoother as inner solver. Inner solves were stopped after 10^{-4} reduction in the residual or at most 10 iterations. Numbers given are average convergence rates after 10^{-4} reduction or 30 iterations.

		nonlinear local multigrid				outer fixed point iteration			
	level	$c=10^1$	$c=10^2$	$c=10^3$	$c=10^4$	$c=10^1$	$c=10^2$	$c=10^3$	$c=10^4$
local	1	$8.93 \cdot 10^{-3}$	$1.23 \cdot 10^{-1}$	$9.24 \cdot 10^{-2}$	$9.47 \cdot 10^{-2}$	$8.79 \cdot 10^{-2}$	$2.91 \cdot 10^{-1}$	$3.04 \cdot 10^{-1}$	$3.06 \cdot 10^{-1}$
grid	2	$2.96 \cdot 10^{-2}$	$7.92 \cdot 10^{-2}$	$7.93 \cdot 10^{-2}$	$8.26 \cdot 10^{-2}$	$6.78 \cdot 10^{-2}$	$2.87 \cdot 10^{-1}$	$2.92 \cdot 10^{-1}$	$3.00 \cdot 10^{-1}$
refinement	3	$2.55 \cdot 10^{-2}$	$1.21 \cdot 10^{-1}$	$2.39 \cdot 10^{-2}$	$2.49 \cdot 10^{-1}$	$7.45 \cdot 10^{-2}$	$3.21 \cdot 10^{-1}$	$3.52 \cdot 10^{-1}$	$3.58 \cdot 10^{-1}$
	4	$2.48 \cdot 10^{-2}$	$5.87 \cdot 10^{-2}$	$2.31 \cdot 10^{-1}$	$2.50 \cdot 10^{-1}$	$6.86 \cdot 10^{-2}$	$3.24 \cdot 10^{-1}$	$3.66 \cdot 10^{-1}$	$3.74 \cdot 10^{-1}$
	5	$2.59 \cdot 10^{-2}$	$3.46 \cdot 10^{-2}$	$2.81 \cdot 10^{-1}$	$2.55 \cdot 10^{-1}$	$3.46 \cdot 10^{-2}$	$3.09 \cdot 10^{-1}$	$3.53 \cdot 10^{-1}$	$3.40 \cdot 10^{-1}$
	6	$3.70 \cdot 10^{-2}$	$3.78 \cdot 10^{-2}$	$3.35 \cdot 10^{-1}$	$2.89 \cdot 10^{-1}$	$2.26 \cdot 10^{-2}$	$3.02 \cdot 10^{-1}$	$3.49 \cdot 10^{-1}$	$3.31 \cdot 10^{-1}$
	7	$3.50 \cdot 10^{-2}$	$3.48 \cdot 10^{-2}$	$3.75 \cdot 10^{-1}$	$3.41 \cdot 10^{-1}$	$2.99 \cdot 10^{-2}$	$3.00 \cdot 10^{-1}$	$3.44 \cdot 10^{-1}$	$3.32 \cdot 10^{-1}$
global	1	$8.93 \cdot 10^{-3}$	$1.23 \cdot 10^{-1}$	$2.56 \cdot 10^{-1}$	$2.56 \cdot 10^{-1}$	$8.79 \cdot 10^{-2}$	$2.91 \cdot 10^{-1}$	$3.04 \cdot 10^{-1}$	$3.06 \cdot 10^{-1}$
grid	2	$2.96 \cdot 10^{-2}$	$7.92 \cdot 10^{-1}$	$2.94 \cdot 10^{-1}$	$2.94 \cdot 10^{-1}$	$6.78 \cdot 10^{-2}$	$2.87 \cdot 10^{-1}$	$2.92 \cdot 10^{-1}$	$3.00 \cdot 10^{-1}$
refinement	3	$3.54 \cdot 10^{-2}$	$9.97 \cdot 10^{-2}$	$3.58 \cdot 10^{-1}$	$3.58 \cdot 10^{-1}$	$8.69 \cdot 10^{-2}$	$3.15 \cdot 10^{-1}$	$3.46 \cdot 10^{-1}$	$3.50 \cdot 10^{-1}$
	4	$3.39 \cdot 10^{-2}$	$5.46 \cdot 10^{-2}$	$3.47 \cdot 10^{-1}$	$3.47 \cdot 10^{-1}$	$7.00 \cdot 10^{-2}$	$2.99 \cdot 10^{-1}$	$3.28 \cdot 10^{-1}$	$3.36 \cdot 10^{-1}$
	5	$3.32 \cdot 10^{-2}$	$3.14 \cdot 10^{-2}$	$3.01 \cdot 10^{-1}$	$3.01 \cdot 10^{-1}$	$2.70 \cdot 10^{-2}$	$2.96 \cdot 10^{-1}$	$2.97 \cdot 10^{-1}$	$3.05 \cdot 10^{-1}$

4   Conclusion

We formulated a multiplicative multigrid algorith with "local" smoothing and applied it to unsymmetric systems obtained from an upwind discretization of the convection-diffusion equation. If a correct ordering of the vertices is used it could be shown that the linear method on unstructured, locally refined grids is as efficient as the methods working on simple structured grids.

For nonlinear problems we formulated an analogue of the nonlinear multigrid method for locally refined grids. The method was applied to a nonlinear convection-diffusion equation and worked well for convection strength below 10^4, but in general a stand alone fixed point iteration with inner linear solves was more stable and more efficient.

Further work will include investigations how effective orderings of vertices for dominating convection can be found automatically. The nonlinear method has to be tested with more problems and also better nonlinear smoothers and damping strategies should be implemented.

References

[1]   R. Bank,T. F. Dupont, H. Yserentant, *The Hierarchical Basis Multigrid Method*, Numer. Math., **52**, 427-458 (1988).

[2]   R. Bank, *PLTMG User's Guide Version 6.0*, SIAM, Philadelphia, 1990.

[3]   P. Bastian, *ug 2.0 Short Manual*, Interdisziplinäres Zentrum für Wissenschaftliches Rechnen, Univ. Heidelberg, 1992.

[4] C. Becker, J.H. Ferziger, M. Peric, and G. Scheuerer, *Finite Volume Multigrid Solution of the Twodimensional, Incompressible Navier-Stokes Equations*, in Robust Multigrid Methods, Proceedings of the Fourth GAMM Seminar, Notes on Numerical Fluid Mechanics, Vieweg Verlag, Braunschweig, **23** (1988).

[5] F. A. Bornemann, *A Sharpened Condition Number Estimate for the BPX Preconditioner of Elleptic Finite Element Problems on Highly Nonuniform Triangulations*, Preprint SC 91-9, Konrad-Zuse-Zentrum, Berlin, 1991.

[6] F.A. Bornemann, H. Yserentant, *A Basic Norm Equivalence for the Theory of Multilevel Methods*, Preprint SC 92-1, Konrad-Zuse-Zentrum, Berlin, 1992

[7] J. H. Bramble, J. E. Pasciak, and J. Xu, *Parallel Multilevel Preconditioners*, Math. Comput., **55**, 1-22 (1990).

[8] J. H. Bramble, J. E. Pasciak, *New Estimates for Multilevel Algorithms Including the V-Cycle*, Technical Report Brookhaven National Laboratory, BNL-46730, 1991.

[9] A. Brandt, *Multi-level Adaptive Solutions to Boundary Value Problems, Mathematics of Computation*, **31**, 333-390(1977).

[10] W. Hackbusch, *Multigrid Methods and Applications*, Springer Verlag, Berlin, Heidelberg, 1985.

[11] W. Hackbusch, *Iterative Lösung großer schwachbesetzter Gleichungssysteme*, Teubner Verlag, Stuttgart, 1991.

[12] W. Hackbusch, *On First and Second Order Box Schemes*, Computing, **41**, 277-296(1989).

[13] M. Hortmann, M. Peric, and G. Scheuerer, *Finite Volume Multigrid Prediction of Laminar Natural Convection: Bench-Mark Solutions*, International Journal for Numerical Methods in Fluids, **11**, 189-207 (1990).

[14] J. A. Meijerink, H. A. van der Vorst, *An Iterative Solution Method for Linear Systems of which the Coefficient Matrix is a Symmetric M-Matrix*, Mathematics of Computation, **31**, 148-162 (1977).

[15] P. Oswald, *On Discrete Norm Estimates Related to Multilevel Preconditioners in the Finite Element Method*, Proc. Int. Conf. Constr. Theory of Functions, Varna, 1991, to appear.

[16] G. E. Schneider, M. J. Raw, *Control Volume Finite-Element Method for Heat Transfer and Fluid Flow Using Colocated Variables − 1. Computational Procedure*, Numerical Heat Transfer, **11**, 363-390(1987).

[17] P. Wesseling, *An Introduction to Multigrid Methods*, John Wiley & Sons, Chichester, 1992.

[18] G. Wittum, *Multigrid Methods for Stokes- and Navier-Stokes Equations*, Numer. Math., **54** (1989).

[19] G. Wittum, *On the Robustness of ILU Smoothing*, SIAM J. Sci. Statist. Comput., **10**, 699-717 (1989).

[20] J. Xu, *Iterative Methods by Space Decomposition and Subspace Correction: A Unifying Approach*, Report No. AM 67, Dep. of Math., Pennsylvania State University, 1990.

[21] H. Yserentant, *On the Multi-Level Splitting of Finite Element Spaces*, Numer. Math., **49**, 379-412 (1986).

[22] H. Yserentant, *Two Preconditioners Based on Multi-Level Splitting of Finite Element Spaces*, Numer. Math., **58**, 163-184 (1990).

INCOMPLETE FACTORIZATIONS WITH S/P AND MODIFIED S/P CONSISTENTLY ORDERED M-FACTORS*

ROBERT BEAUWENS
SERVICE DE MÉTROLOGIE NUCLÉAIRE
UNIVERSITÉ LIBRE DE BRUXELLES, C.P. 165
50, AV. F.D. ROOSEVELT, B-1050 BRUSSELS.

Summary

We report here on the first results obtained in the development of a new approximate factorization technique based on the properties of "S/P consistently ordered" M-matrices. An ordered undirected graph is called S/P consistently ordered if whenever a pair of nodes $i < j$ has a common neighbour k with $k < i$, then either it has also a common neighbour ℓ with $j < \ell$ or neither i has any neighbour ℓ with $\ell > i$ nor j any neighbour ℓ with $\ell > j$. Incomplete factorizations with (M-matrix) triangular factors whose graphs are S/P consistently ordered have attractive properties which have been analysed. The conclusion of our analysis is that in such cases, the submatrix determined by the nodes without successors should be factored exactly.

1. Introduction

We review here some recent advances in the analysis of the conditioning of Stieltjes matrices by unperturbed modified incomplete factorizations obtained in [7, 4, 15] under the assumption that the approximate triangular factors are "S/P consistently ordered" M-matrices. The improvements obtained in these works bear on a definition of the notion of "S/P consistency" more general than the one formerly introduced in [6] and show that, under these conditions, the submatrix determined by the set of nodes without successors should be modified to an exact factorization.

Besides fully generalizing the notion of "S/P consistency", this new definition has the additional property that *any* consistently ordered graph can now be graph perturbed (i.e. by addition of new edges) to an S/P consistently ordered one [8]. Since consistent orderings represent a general class of orderings suitable for parallel or vector implementations, our hope is that it will open the way to the development of appropriately controlled modified approximate factorizations of Stieltjes matrices arising from discrete PDE's under grid orderings conveniently tailored to such implementations, a domain where rather negative results have been reported [10].

The modification reported here represents the very first step of this new way of controlling the associated modified approximate factorizations, and often the only one

---

* The present work was supported by the "Programme d'impulsion en Technologie de l'Information" financed by Belgian State, under contract Nr. IT/IF/14.

to be taken in the particular case where they remain *incomplete* factorizations. As a matter of fact, we recover the good behaviour usually expected under the standard lexicographic ordering on rectangular grids. Further, since this ordering is precisely our first example of an S/P consistent ordering, its good behaviour may plainly be ascribed to this specific feature.

2. General terminology and notation

All graphs considered here are ordered undirected graphs with node set $[1,n]$, i.e. the ordered set of the n first integers; all matrices are real $n \times n$ matrices; all vectors are real n-vectors.

Neighbours j of a node i in a graph G with $j < i$ $(j > i)$ are called precursors (successors); the set of neighbours (precursors, successors) of i is denoted by $N(i)(P(i), S(i))$; for $M \subset [1, n]$, we let

$$N(M) = \bigcup_{i \in M} N(i) , \ P(M) = \bigcup_{i \in M} P(i) , \ S(M) = \bigcup_{I \in M} S(i) ; \tag{2.1}$$

for any pair of nodes, we denote by

$$Pc(i,j) = P(i) \bigcap P(j) \quad (Sc(i,j) = S(i) \bigcap S(j)) \tag{2.2}$$

their set of common precursors (successors) and we let

$$Pc(G) = \bigcup_{i \neq j} Pc(i,j) \quad Sc(G) = \bigcup_{i \neq j} Sc(i,j) . \tag{2.3}$$

If G is the graph $G(A)$ of the matrix A, we also use $Pc(A)$ and $Sc(A)$ for $Pc(G)$ and $Sc(G)$.

3. S/P consistently ordered graphs

According to [7], a graph is said S/P consistently ordered if, for any pair of nodes i, j with $i \neq j$,

$$Pc(i,j) \neq \emptyset \ \Rightarrow \ Sc(i,j) \neq \emptyset \ \text{or} \ S(i) = S(j) = \emptyset . \tag{3.1}$$

Figures 1 to 8 give a few specific examples to illustrate this definition (in all cases, the nodes without successors are represented by small empty circles).

Figure 1 shows that the lexicographic ordering of a cartesian grid on a rectangle produces an S/P consistently ordered graph while Figure 2 shows that the same ordering may fail to do so on an L-shaped domain. The S/P consistency assumption may however be recovered in this example by reversing the ordering or by adding the edge {11,20}. The addition of the edge {12,16} could also be considered, but it requires first to renumber the nodes 12,13,...,18,19 as 19,12,...,17,18 (for example) and then to add the edge {15,19}.

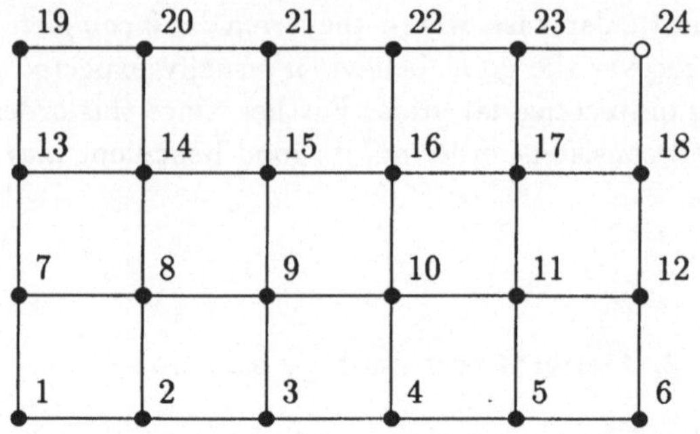

FIG. 1. *Lexicographic ordering of a cartesian grid on a rectangle.*

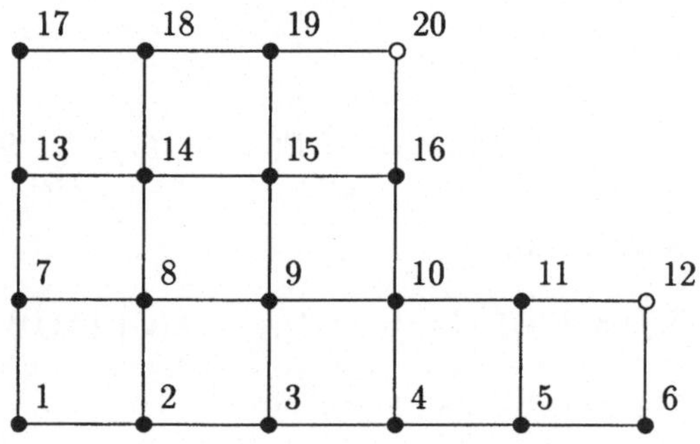

FIG. 2. *Lexicographic ordering of a cartesian grid on an L-shaped domain.*

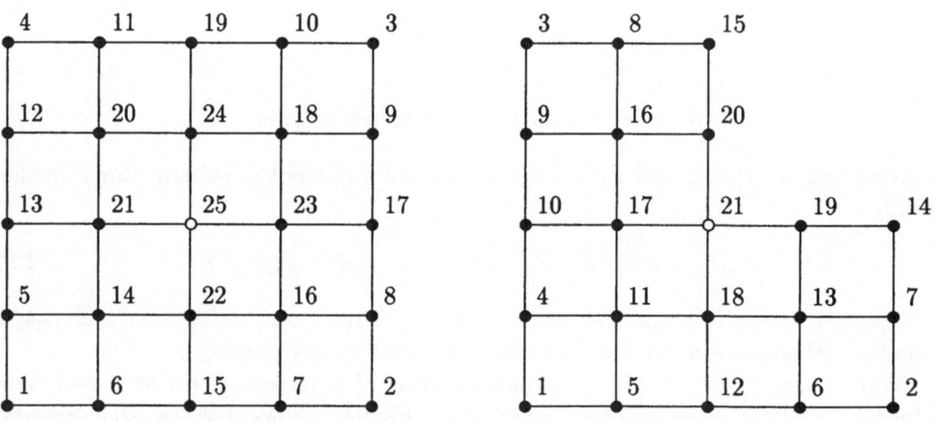

FIG. 3. *Notay's ordering of a cartesian grid on a square and on an L-shaped domain.*

Figure 3 shows cartesian grids ordered by Notay's ordering which consists in choosing arbitrarily the last node and then applying recursively the rule : number next unnumbered neighbours as next precursors. Notay's claim is that any cartesian grid (without holes) numbered in this way produces an S/P consistently ordered graph.

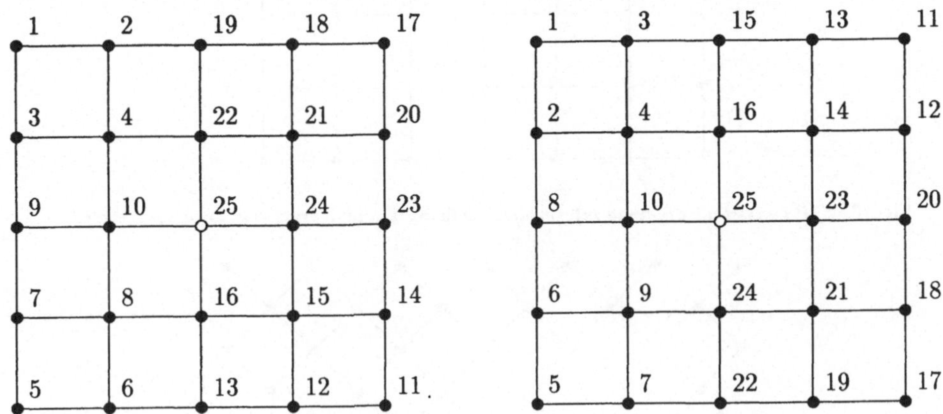

FIG. 4. *Two versions [10] of van der Vorst's 4-processor ordering on a square grid.*

Notay's ordering may be considered as a variant of the reverse Cuthill-McKee ordering (from which it differs however since reversing Notay's ordering need not reproduce the Cuthill-McKee ordering). For two-dimensional cartesian grids, it may also be viewed as a variant of the van der Vorst 4-processor ordering, illustrated on Figure 4 (both version of which are S/P consistent orderings).

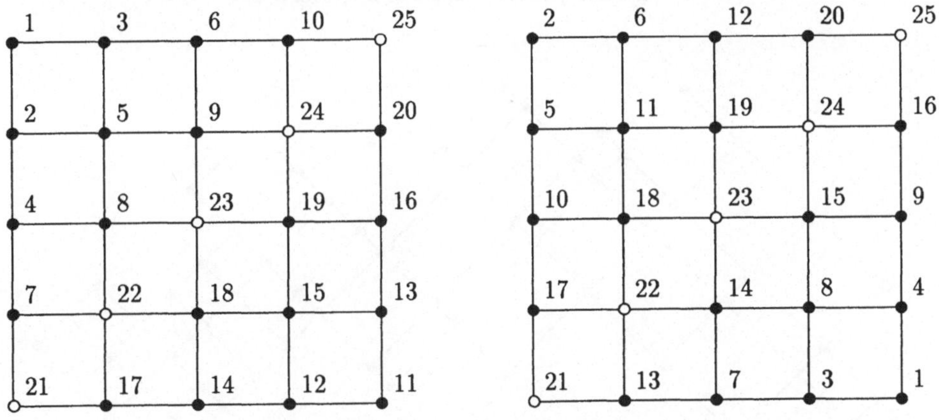

FIG. 5. *Eijkhout's 2-processor ordering on a square grid & its variant obtained by Notay's rule.*

Figure 5 (left) illustrates Eijkhout's 2-processor ordering of a square grid [11] which is S/P consistent according to the new definition although not according to the former one (which did not allow more than one node without successor per connected component). Notice that essentially the same ordering may also be obtained by applying Notay's rule after having chosen and numbered as last nodes all nodes of the main diagonal of the square, as also shown on Figure 5 (right).

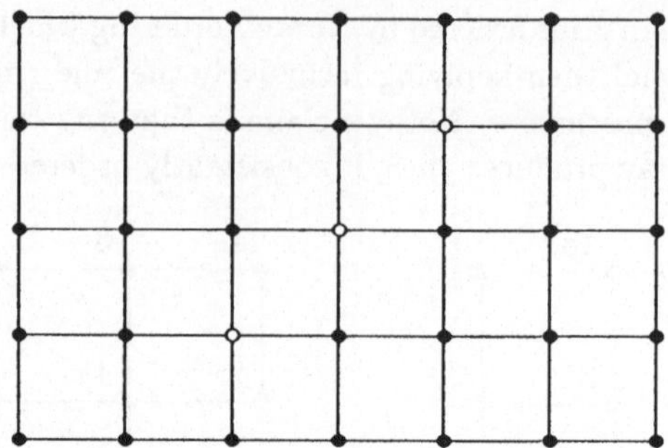

FIG. 6. *S/P consistent ordering intermediate between 2 and 4-processor on a square grid.*

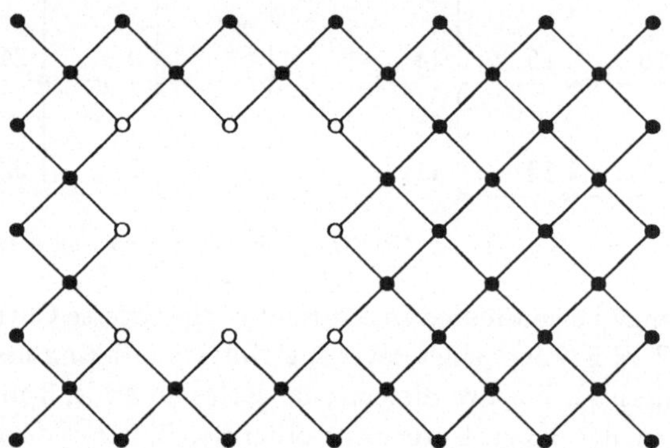

FIG. 7. *Cartesian grid diagonally oriented, with a hole.*

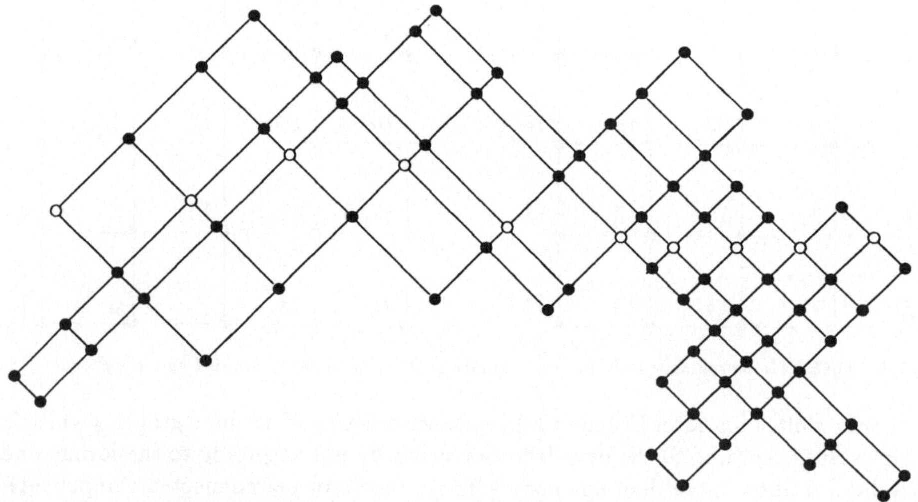

FIG. 8. *Irregular cartesian grid on an irregular region.*

The last remark leads to various generalizations obtained in the same way, which are illustrated on Figures 6 to 8. Node numbers are not indicated since they are generated, as explained, by Notay's rule after having chosen and numbered the nodes without successors as last nodes (empty circles). The ordering of Figure 6 is intermediate between the 2-processor and the 4-processor orderings. Figure 7 shows a (diagonally oriented) cartesian grid on a region with a hole. Figure 8 shows an irregular cartesian grid on an irregular region.

Further generalizations involve 3-dimensional cartesian meshes which can similarly be S/P consistently ordered by following Notay's rule once an appropriate set of nodes without successors has been chosen.

4. Consistent orderings

A graph is consistently ordered if there exists a partitioning $\mathcal{L} = (L_k)_{0 \leq k \leq \ell}$ of its node set such that

$$P(L_k) \subset L_{k-1} \tag{4.1}$$

for $0 \leq k \leq \ell$ (being understood that $L_{-1} = \emptyset$).

Note that :

(1) the partitioning $\mathcal{L}$ which occurs here is a particular example of a "level structure"(cf. [12]);

(2) if a graph is consistently ordered and connected, then $\mathcal{L}$ is unique (cf. [5]);

(3) consistent orderings are suitable for parallel and vector implementations because the subgraph determined by the nodes of any level L_k is totally disconnected;

(4) such level structures are readily identified in all the examples of the preceding section.

The last remark leads to :

THEOREM 4.1. [7] *An S/P consistently ordered graph is consistently ordered.*

We further claim that :

THEOREM 4.2. [8] *If G is consistently ordered, then there exists (possibly after some reordering of G) an ordered graph G' such that $G \subset G'$.*

An example was discussed above showing why some reordering might first be required.

5. Quantifying the notion of "S/P consistency"

Let us first recall that the notion of diagonal dominance may be defined both as a qualitative and as a quantitative notion, as illustrated by the following definition.

DEFINITION 5.1. *Let $A = (a_{ij})$ be a matrix with positive diagonal entries and non-positive offdiagonal entries, let x be a positive vector and let $M \subset [1, n]$. We call dominance ratio of A with respect to M and x the number*

$$t = max_{i \in M}\left(\frac{-\sum_{j \neq i} a_{ij} x_j}{a_{ii} x_i}\right) \tag{5.1}$$

with $t = 0$ if $M = \emptyset$. A is diagonally dominant (with respect to M and x) if $t \leq 1$ and strictly diagonally dominant (with respect to M and x) if $t < 1$.

The notion of "S/P consistency" can also be quantified, by means of the "maximal reduction ration"as defined below (see also [7] and [15]).

DEFINITION 5.2. *Let $U = P - F$ be an upper triangular M-matrix with $P = diag(U)$ and let F' be defined by*

$$f'_{ij} = \begin{cases} f_{ij} & \text{if } S(S(i)) \neq \emptyset \\ 0 & \text{otherwise} \end{cases} \qquad (5.2)$$

where f_{ij} and f'_{ij} denote the entries of F and F' respectively. Then, a lower triangular M-matrix $L = Q - E$ with $Q = diag(L)$ is an S/P image of U if

$$E \leq F^T \qquad (5.3)$$

with

$$offdiag(E^T Q^{-1} E) \geq offdiag(F'^T P^{-1} F') \ . \qquad (5.4)$$

A reduced S/P image of U is an S/P image of KU where $K = (\kappa_i \delta_{ij})$ with $0 < \kappa_i \leq 1$ for all $1 \leq i \leq n$. K is the reduction matrix and $\kappa = min_i \kappa_i$ is the reduction ratio.

DEFINITION 5.3. *Let U be an upper triangular S/P consistently ordered M-matrix, x a positive vector and t a positive number. We call maximum reduction ratio of U with respect to x and t and we denote by $\kappa_t(U, x)$ the maximal value of κ such that U has a reduced S/P image L of reduction ratio κ and whose dominance ratio with respect to $Sc(L)$ and x does not exceed t.*

6. Modified incomplete factorizations

Let A be a Stieltjes matrix and U an upper triangular M-matrix with $P = diag(U)$ such that

$$offdiag(A - (U^T - P)P^{-1}(U - P)) \leq offdiag(U^T + U) \leq offdiag(A) \ . \qquad (6.1)$$

Let x be some positive vector such that $Ax \geq 0$ and let t be the dominance ratio of U with respect to $Pc(U)$ and x.

Letting $B = U^T P^{-1} U$, notice that the assumption (6.1) is satisfied a.o. when B is any incomplete factorization of A.

If, further,

$$t \leq \tau < 1 \qquad (6.2)$$

and

$$(1 - \tau)Ax \leq Bx \leq Ax \ , \qquad (6.3)$$

then, one has [9] the following upper bound on the spectral condition number of $B^{-1}A$

$$\kappa(B^{-1}A) \leq \frac{1}{1 - \tau} \ . \qquad (6.4)$$

Notice that the best possible choice for Bx within the constraints (6.3) is (cf. [9])

$$Bx = Ax \tag{6.5}$$

i.e. the rule which defines the so-called unperturbed modified factorizations (N.B. This is the best possible choice with respect to *nonpositive* perturbations since positive ones are forbidden by the assumption (6.3)).

This result tells that the measure of the strict diagonal dominance of U with respect to $Pc(U)$ and x as given by $\alpha = 1 - \tau$ directly controls the upper bound $1/\alpha$ on $\kappa(B^{-1}A)$.

This bound has been very useful in the past –under a variety of geometrical disguises– but it has the shortcoming of disappearing for $t = 1$, i.e. when U is only weakly diagonally dominant with respect to $Pc(U)$ and x, leaving us with a generally unpredictible value of $\kappa(B^{-1}A)$.

However, if U is S/P consistently ordered with associated level structure $\mathcal{L} = (L_k)_{0 \geq k \geq \ell}$, if we modify $U = (u_{ij})$ into $\tilde{U} = (\tilde{u}_{ij})$ defined by

$$\begin{aligned}
\tilde{u}_{ij} &= u_{ij} && \text{for } i \notin L_\ell \\
\tilde{u}_{ij} &= a_{ij} - \sum_{k<i} \frac{\tilde{u}_{ki}\tilde{u}_{kj}}{\tilde{u}_{kk}} && \text{for } i \in L_\ell
\end{aligned} \tag{6.6}$$

and let $\tilde{B} = \tilde{U}^T \tilde{P}^{-1} \tilde{U}$ with $\tilde{P} = diag(\tilde{U})$, then exchanging the assumption (6.2) for

$$\tilde{t} \leq \tau \leq 1 \tag{6.7}$$

where $\tilde{t}$ is the dominance ratio of U with respect to $Pc(U) \cap (\bigcup_{k=0}^{\ell-2} L_k)$ and x, and exchanging the assumption (6.3) for

$$(1 - \tau + \frac{\kappa}{\ell+2}\tau)Ax \;\leq\; \tilde{B}x \;\leq\; Ax \tag{6.8}$$

with $\kappa \geq \kappa_{\tilde{t}}(U,x)$, it follows from Theorem 5.1 of [4] that the spectral condition number of $\tilde{B}^{-1}A$ is bounded by

$$\kappa(\tilde{B}^{-1}A) \;\leq\; \frac{1}{1 - \tau + \frac{\kappa}{\ell+2}\tau} \tag{6.9}$$

which never disappears since it implies

$$\kappa(\tilde{B}^{-1}A) \;\leq\; \frac{\ell+2}{\kappa} \; . \tag{6.10}$$

This new result tells that the measure of diagonal dominance of U with respect to $Pc(U) \cap (\bigcup_{k=0}^{\ell-2} L_k)$ as given by $\alpha_d = 1 - \tau$ (possibly zero) and the measure of S/P consistency of U as given by $\alpha_c = \frac{\kappa}{\ell+2}$ essentially sum up their values to produce the upper bound $1/(\alpha_d + \tau\alpha_c)$ on $\kappa(\tilde{B}^{-1}A)$.

This result obviously recommends the use of $\tilde{B}$ rather than B and we shall now review relevant issues associated with this modification. We restrict our discussion to the case where $G(U)$ is connected (for otherwise the same comments should be applied to each connected component of $G(U)$ separately).

The first relevant question is the existence of B. It follows from Notay's criterion [16] that B is unsafe whenever the graph of U has more than one node without successors and appropriate warnings have been raised by Eijkhout [11] against blind use of unperturbed modified factorizations in such cases, most often arising from the introduction of non standard orderings, as for example in [10]. The modification proposed here solves this problem because the fill-in introduced in $\widetilde{U}$ reduces the number of nodes without successors to one. Of course, S/P consistency is an excessive requirement to reach this purpose. All what is needed is that the set of nodes where all fill-in is accepted, call it N, should include all nodes without successors as well as a sufficient number of precursors (of other members of N) in such a way that the subgraph induced by $N \cup P(N)$ in $G(U)$ should be connected. It is appropriate to note here that with respect to S/P consistency, the results of [4] do not require that $G(U)$ be S/P consistently ordered, but only that the subgraph induced by $M = [1, n] \backslash N$ in $G(U)$ should be S/P consistently ordered.

These comments show that a key issue with respect to ordering is the appropriate choice of the set N. Besides existence (of $\widetilde{B}$) and S/P consistency (of the subgraph induced by M) one should also try to have the size of N reasonably small to avoid excessive fill-in. This consideration makes it impossible for example to apply our recommendation to red-block orderings of cartesian grids (despite the fact that such orderings would otherwise be ideally suited to our needs). At the other extreme, there may be no distinction between B and $\widetilde{B}$. This is the particular situation that had already been analysed in [6] with the former definition of S/P consistency.

The next question concerns the bounds on $\kappa(\widetilde{B}^{-1}A)$ and $\kappa(B^{-1}A)$. Provided that the subgraph induced in $G(U)$ by the set M defined above is S/P consistent, we always have a predictible upper bound on $\kappa(\widetilde{B}^{-1}A)$, but generally not on $\kappa(B^{-1}A)$. Further, whenever both exist (and $Card(N) > 1$), the bound on the former is always smaller than the bound on the latter. Numerical experiments reported in [4] show that the actual value of $\kappa(\widetilde{B}^{-1}A)$ is indeed smaller than $\kappa(B^{-1}A)$ and that the reduction may range from moderate to very large factors.

That the S/P consistency assumption is required for the last results, at least in present stage of knowledge, may be illustrated by Theorem 3.1 of [4] where a bound on $\kappa(\widetilde{B}^{-1}A)$ is derived without the S/P consistency assumption. This bound is again of the form $1/(1 - \tau)$ with only a slightly improved value of τ and it also dies out when $\tau = 1$. Other bounds, derived by Axelsson [2] and Notay [17] were found inapplicable to the numerical examples of [4] because the matrix $U + U^T - A$ was not positive definite.

The last issue to be mentioned is when and how one can control the basic parameters τ and $\kappa_t(U, x)$ of the method. The use of perturbations acting on τ has already a long history since it is the device introduced and developed by Axelsson and his coworkers [1, 13, 14, 3] under lexicographic ordering on cartesian grids; and it was recently successfully used by Eijkhout [11] with his 2-processor ordering on a square grid. The use of perturbations acting on $\kappa_t(U, x)$ is under investigation [8]. In both cases, the lower end of the spectrum of $\widetilde{B}^{-1}A$ is perturbed and has to be estimated.

REFERENCES

[1] O. AXELSSON, *A generalized SSOR method*, BIT, 13 (1972), pp. 443–467.

[2] ———, *On iterative solution of elliptic difference equations on a mesh connected array of processors*, J. High Speed Comput., 1 (1989), pp. 165–184.

[3] O. AXELSSON AND V. BARKER, *Finite Element Solution of Boundary Value Problems. Theory and Computation*, Academic Press, New York, 1984.

[4] R. BEAUWENS, *Approximate factorizations with modified S/P consistently ordered M-factors*, Numer. Lin. Alg. with Appl., submitted.

[5] ———, *Consistent ordering analysis*, Scientific Report, Université Libre de Bruxelles, Brussels, 1986.

[6] ———, *Approximate factorizations with S/P consistently ordered M-factors*, BIT, 29 (1989), pp. 658–681.

[7] R. BEAUWENS, Y. NOTAY, AND B. TOMBUYSES, *S/P images of upper triangular M-matrices*, Numer. Lin. Alg. with Appl., submitted.

[8] R. BEAUWENS AND B. TOMBUYSES, *Graph perturbations*, in progress.

[9] R. BEAUWENS AND R. WILMET, *Conditioning analysis of positive definite matrices by approximate factorizations*, J. Comput. Appl. Math., 26 (1989), pp. 257–269.

[10] I. DUFF AND G. MEURANT, *The effect of ordering on preconditioned conjugate gradients*, BIT, 29 (1989), pp. 635–657.

[11] V. EIJKHOUT, *Beware of unperturbed modified incomplete factorizations*, in Iterative Methods in Linear Algebra, R. Beauwens and P. de Groen, eds., North-Holland, 1992.

[12] A. GEORGE AND J. LIU, *Computer solution of large sparse positive definite systems*, Prentice-Hall, Englewood Cliffs, 1981.

[13] I. GUSTAFSSON, *A class of first order factorization methods*, BIT, 18 (1978), pp. 142–156.

[14] ———, *Modified Incomplete Cholesky (MIC) methods*, in Preconditioning Methods. Theory and Applications, D. Evans, ed., Gordon and Breach, New York-London-Paris, 1983, pp. 265–293.

[15] Y. NOTAY, *Conditioning of Stieltjes matrices by S/P consistently ordered approximate factorizations*, Appl. Numer. Math., to appear.

[16] ———, *Solving positive (semi)definite linear systems by preconditioned iterative methods*, in Preconditioned Conjugate Gradient Methods, O. Axelsson and L. Kolotilina, eds., Lectures Notes in Mathematics No. 1457, Springer-Verlag, 1990, pp. 105–125.

[17] ———, *Upper eigenvalue bounds and related modified incomplete factorization strategies*, in Iterative Methods in Linear Algebra, R. Beauwens and P. de Groen, eds., North-Holland, 1992, pp. 551–562.

GRID- AND POINT-ORIENTED MULTILEVEL ALGORITHMS

MICHAEL GRIEBEL

Institut für Informatik, Technische Universität München
Arcisstraße 21, D-8000 München 2
email: griebel@informatik.tu-muenchen.de

Abstract

Instead of the usual basis, we use a generating system for the discretization of PDEs that contains not only the basis functions of the finest level of discretization but additionally the basis functions of all coarser levels of discretization. The Galerkin-approach now results in a semidefinite system of linear equations to be solved. Standard iterative GS-methods for this system turn out to be equivalent to elaborated multigrid methods for the fine grid system.

Beside Gauss-Seidel methods for the level-wise ordered semidefinite system, we study block Gauss-Seidel methods for the point-wise ordered semidefinite system. These new algorithms show the same properties as conventional multigrid methods with respect to their convergence behavior and efficiency. Additionally, they possess superior properties with respect to parallelization.

We report the results of numerical experiments regarding the reduction rates of different variants of these algorithms.

Key words. partial differential equations, multilevel methods, multigrid methods, Gauss-Seidel iteration, block Gauss-Seidel iteration, semidefinite system.

AMS(MOS) subject classifications. 65F10, 65N20, 65N30

1. Introduction

Recently, see [5], a new concept for the development of multigrid and BPX-like multilevel algorithms for standard full grid problems has been presented. There, instead of a basis approach on the finest grid and the acceleration of the basic iteration by MG-coarse grid correction or a BPX-type preconditioner, a generating system is used to allow a non-unique level-wise decomposed representation of the solution. The degrees of freedom are associated to the nodal basis functions of all levels under consideration. With this non-unique multilevel decomposed representation of a function, the Galerkin-approach leads to a semidefinite linear system with unknowns on all levels. Its solution is non-unique but in some sense equivalent to the unique solution of the standard problem on the finest grid.

Furthermore, it has been shown that traditional iterative methods for the semidefinite system are equivalent to modern elaborated multilevel methods applied to the standard system which exhibit optimal convergence properties. The conjugate gradient method (with appropriate diagonal scaling) for the semidefinite system is equivalent to the BPX-conjugate gradient method for the fine grid system. Gauss-Seidel-type iterations for the semidefinite system are equivalent to certain multigrid methods. For details, see [5]. These methods are grid- or level-oriented and can be considered as level-block techniques. An outer iteration switches from level to level, and an inner iteration operates on the specific grid.

Now, we consider the semidefinite system from a different point of view. We group all unknowns together which are associated to the same grid point. This results in a point-oriented method and can be seen as a point-block technique. Now, an outer iteration switches from grid point to grid point. The local system that belongs to all basis functions of different levels centered in the same grid point can be solved either directly or by an inner iteration that runs over all levels that are associated to the grid point under consideration. Furthermore, grid points can be grouped together to form subdomains. In this sense, we get some sort of simple domain decomposition method which exhibits MG-like convergence properties.

In contrast to the parallelization of a multilevel method where communication has to take place on all levels to maintain good convergence rates, our point-block approach needs substantially less communication due to its domain decomposition qualities. In this sense, our new method is superior to other parallel multigrid and multilevel methods and we believe that its future will be bright.

Note that our point-oriented methods are closely related to the frequency filtering ILU technique [14]. Instead of discrete *sinus* functions for the filtering, we use the frequency components that belong to the same grid point but consider them with respect to different mesh sizes.

We report the results of numerical experiments regarding the reduction rates of these new algorithms.

2. The semidefinite system

In this section, we introduce a generating system that replaces the usual finite element basis in the discretization of a boundary value problem of a partial differential equation. We then derive the associated semidefinite linear system and discuss its properties.

Consider a partial differential equation in d dimensions with a linear, second-order operator L on the domain $\Omega = (0,1)^d$, $d=1,2,...,$

$$Lu = f \text{ on } \Omega, \tag{1}$$

with appropriate boundary conditions and corresponding solution u. For reasons of simplicity, we restrict ourselves to homogeneous Dirichlet boundary conditions. Given an appropriate function space V, the problem can be expressed equivalently in its variational form: Find a function $u \in V$ with

$$a(u,v) = (f,v) \quad \forall\, v \in V. \tag{2}$$

(In the case of homogeneous Dirichlet boundary conditions, V would be the Sobolev space $H_0^1(\Omega)$.) Here, $a\colon V \times V \to \mathbb{R}$ is a bounded, positive-definite, symmetric bilinear form and $(.,.)$ is the linear form for the right-hand side. Let $\|.\|_a := \sqrt{a(.,.)}$ denote the induced energy norm. We assume that V is complete with respect to $\|.\|_a$, which is true if $a(.,.)$ is H_0^1-elliptic. The Lax-Milgram lemma then guarantees the existence and uniqueness of the solution of (2). If we consider directly the functional $J(u) :=$

$1/2 \cdot a(u,u) - (f,u)$, the problem can be stated alternatively as minimization of $J(u)$ in V.

2.1. Spaces, bases, and the generating system. Assume we are given a sequence of uniform, equidistant, and nested grids

$$\Omega_1 \subset \Omega_2 \subset ... \subset \Omega_{k-1} \subset \Omega_k \tag{3}$$

on $\bar{\Omega}$ with respective mesh sizes $h_l = 2^{-l}$, $l=1,..,k$, and an associated sequence of spaces V_l of piecewise d-linear functions,

$$V_1 \subset V_2 \subset .... \subset V_{k-1} \subset V_k, \tag{4}$$

with dimensions

$$n_l := dim(V_l) = (2^l - 1)^d, l = 1,..,k. \tag{5}$$

Here, 1 denotes the coarsest and k the finest level of discretization. Consider also the sequence

$$N_1 \subset N_2 \subset ... \subset N_{k-1} \subset N_k \tag{6}$$

of sets of inner grid points $N_l = \{x_1, ..., x_{n_l}\}$ of the grid Ω_l, $l=1,..,k$.

The standard finite element basis, that spans V_l on the equidistant grid Ω_l, is denoted by B_l. It contains the nodal basis functions $\phi_i^{(l)}$, $i=1,..,n_l$, which are defined by

$$\phi_i^{(l)}(x_j) = \delta_{i,j}, \quad x_j \in N_l. \tag{7}$$

Note that we use rectangular grids. Therefore, the 2D basis functions, e.g., can be written as the product of two 1D basis functions for the different coordinate directions.

Any function $u \in V_k$ can be expressed uniquely by

$$u = \sum_{\phi \in B_k} u_\phi \cdot \phi \tag{8}$$

with the vector $u_k := (u_\phi)_{\phi \in B_k}$ of nodal values associated to some given ordering of the functions of B_k.

Following [16], a hierarchical basis can be constructed, that takes into account all those basis functions from B_l that are not contained in $B_{l-1}, l = k,..,2$. With

$$\tilde{B}_l = \begin{cases} B_1 & \text{for } l = 1, \\ \{\phi_i^{(l)} \in B_l : x_i \notin N_{l-1}\} & \text{for } l > 1, \end{cases} \tag{9}$$

we obtain the hierarchical basis

$$H_k = \bigcup_{l=1}^{k} \tilde{B}_l. \tag{10}$$

Now, a function $u \in V_k$ can be expressed uniquely by

$$u = \sum_{\phi \in H_k} v_\phi \cdot \phi \tag{11}$$

with the vector $v_k := (v_\phi)_{\phi \in H_k}$ of hierarchical values associated to some given ordering of the functions of H_k.

In contrast to these basis approaches expressed by the representations in (8) and (11), we now consider the set of functions E_k defined as the union of all the different nodal bases B_l for the levels $l=1,..,k$,

$$E_k = B_1 \cup B_2 \cup .... \cup B_{k-1} \cup B_k. \tag{12}$$

Obviously, being a linearly dependent set of functions, E_k is no longer a basis for V_k, but merely a generating system. See Figure 1 for a simple 1D example.

FIG. 1. *Functions of the generating system E_3 in the 1D case.*

In any case, an arbitrary function $u \in V_k$ can be expressed in terms of the generating system by

$$u = \sum_{\phi \in E_k} w_\phi \cdot \phi \tag{13}$$

with the vector $w_k^E := (w_\phi)_{\phi \in E_k}$ associated to some given ordering of the functions of E_k.

Here and in the following, we denote representations in terms of the generating system E_k by the superscript E. The length of w_k^E is

$$n_k^E = \sum_{l=1}^{k} n_l, \tag{14}$$

which is in the 1D case about twice, in the 2D case about 4/3 times, and in the 3D case about 8/7 times as large as the length of the vectors for the basis representation above. This is due to the geometric rate of decrease of the number of grid points from fine to coarse levels with factors of 1/2, 1/4, and 1/8 for 1D, 2D, and 3D, respectively. The generalization of this concept to higher dimensions is straightforward.

Note that the representation of u in terms of E_k is not unique. In general, there exists an n_{k-1}^E-dimensional variety of level-wise decompositions of $u \in V_k$. However, for a given representation w_k^E of u in E_k, we can easily compute its unique representation u_k with respect to B_k. This involves in 2D the bilinear interpolation which can be expressed and implemented by MG-prolongation operators.

2.2. Galerkin-approach and linear systems. Using the nodal basis B_k, the Galerkin-approach results in the discrete variational problem for $u \in V_k$

$$\forall \phi_i \in B_k : a(u, \phi_i) = f(\phi_i) \tag{15}$$

and the equivalent linear system of equations for the vector of nodal values

$$L_k u_k = f_k, \tag{16}$$

where

$$\begin{aligned}
(L_k)_{i,j} &:= a(\phi_j^{(k)}, \phi_i^{(k)}), & 1 \le i,j \le n_k \\
(f_k)_i &:= (f, \phi_i^{(k)}), & 1 \le i \le n_k
\end{aligned} \tag{17}$$

for some appropriate ordering of the functions of B_k.

For the generating system E_k, the Galerkin-approach leads to the variational problem

$$\forall \phi_i \in E_k : a(u, \phi_i) = f(\phi_i) \tag{18}$$

and, with representation (13), to the linear system

$$L_k^E w_k^E = f_k^E, \tag{19}$$

where

$$\begin{aligned}
(L_k^E)_{i,j} &:= a(\phi_j, \phi_i), & 1 \le i,j \le n_k^E \\
(f_k^E)_i &:= (f, \phi_i), & 1 \le i \le n_k^E
\end{aligned} \tag{20}$$

for some appropriate ordering of the functions of E_k.

The system $L_k^E w_k^E = f_k^E$ has the following properties. The matrix L_k^E is semidefinite and has the same rank as L_k. Thus $n_{k-1}^E = n_k^E - rank(L_k)$ eigenvalues of L_k^E are zero. The system is solvable because the right-hand side is constructed in a consistent manner. It has not just one unique solution, but a variety of different solutions. However, the evaluation of two different solutions $w_k^{E,1}$ and $w_k^{E,2}$ with respect to their representation in B_k by means of MG-prolongation operators results in the unique solution u_k of the system $L_k u_k = f_k$. Therefore, it is sufficient to compute just one solution of the enlarged semidefinite system to obtain, via interpolation, the unique solution of the system $L_k u_k = f_k$.

Note that the enlarged matrix L_k^E contains the submatrices L_l that arise from the use of the standard basis B_l, $l = 1, .., k$. A similar property holds for the right-hand sides. The matrix and right-hand side of the system

$$\hat{L}_k \hat{u}_k = \hat{f}_k, \tag{21}$$

where

$$(\hat{L}_k)_{i,j} \;:=\; a(\phi_j, \phi_i), \qquad \phi_i, \phi_j \in H_k$$
$$(\hat{f}_k)_i \;:=\; (f, \phi_i), \qquad \phi_i \in H_k, \tag{22}$$

that stems from the hierarchical basis discretization are also contained in L_k^E and f_k^E.

3. Iterative Methods

Now, multilevel-type algorithms are easy to construct. In [5], it is shown that (after appropriate scaling) the conjugate gradient method for the semidefinite system $L_k^E w_k^E = f_k^E$ is equivalent to CG with the preconditioner of Bramble, Pasciak, and Xu [2] for the system $L_k u_k = f_k$. The Gauss-Seidel iteration for the semidefinite system is equivalent to the multigrid-method with Gauss-Seidel smoother applied to $L_k u_k = f_k$. Here, depending on the ordering of the unknowns of the semidefinite system, different MG-cycle strategies can be modeled easily. See the discussion in [5] for further guidance.

In the following, we focus on Gauss-Seidel-type iterations for the semidefinite system only. First, we consider level-oriented Gauss-Seidel iterations and demonstrate that they are equivalent to multigrid and hierarchical basis multigrid methods. An outer iteration switches from level to level, and an inner iteration operates on the respective grids. Then, we study block-oriented Gauss-Seidel iterations. There, the unknowns of the semidefinite system that belong to basis functions centered at the same grid point are grouped together. Now, an outer iteration switches from grid point to grid point, and an inner iteration or a direct solver works on the local subsystem of the semidefinite problem belonging to the respective grid point.

Note that, in both cases, it is not necessary to assemble the matrix L_k^E and the right-hand side f_k^E explicitly. It is possible to use MG-prolongation and MG-restriction operators and the fine grid discretizations L_k, f_k to express L_k^E and f_k^E in a certain product form. For an example for the level-oriented approach, see (35) in [5].

Thus, for appropriate traversal orderings, the level- and point-oriented GS-methods can be implemented to need $O(n_k^E) = O(n_k)$ operations per iteration step, only. Especially for the point-oriented block GS, this is quite tricky. A detailed description of implementation details will be given in a separate paper. The basic approach follows the idea presented in [17] and [3]. We use binary trees (of binary trees) to implement the fine grid. Altogether, the required storage and the number of operations to perform a GS- or block GS-step is proportional to the number of grid points employed.

3.1. Grid-oriented Gauss-Seidel algorithms. We consider Gauss-Seidel iterations for the semidefinite system $L_k^E w_k^E = f_k^E$. As usual, we decompose the semidefinite matrix L_k^E by $L_k^E = F_k^E + G_k^E + (G_k^E)^T$, where F_k^E and G_k^E denote the diagonal and strictly lower triangular parts of L_k^E, respectively. Then, the Gauss-Seidel iteration (GS) is expressed by using $C_k^{E,GS} := (F_k^E + G_k^E)^{-1}$, and its symmetric counterpart SGS is expressed by using $C_k^{E,SGS} := (F_k^E + G_k^E)^{-T} F_k^E (F_k^E + G_k^E)^{-1}$ in the iteration

$$w_k^{E,(\kappa+1)} := w_k^{E,(\kappa)} + C_k^E (f_k^E - L_k^E w_k^{E,(\kappa)}). \tag{23}$$

Note that $F_k^E + G_k^E$ is generally not symmetric, but it is positive definite and invertible.

Of course, the lower triangular part of L_k^E depends on the ordering of the unknowns. For practical reasons, i.e. to maintain $O(n_k)$ operations per iteration step, not all orderings are advisable. Here, we order the unknowns grid-wise (starting for example with the finest grid). This leads to a grid-oriented block partition of L_k^E. The unknowns of each level can be ordered lexicographically or, in the 2D case, in a four color manner.

Note that this level-wise ordering corresponds to a level-wise decomposition

$$V_k = \sum_{l=1}^{k} V_l = \sum_{l=1}^{k} \sum_{i=1}^{n_l} V_{l,x_i},\tag{24}$$

of V_k, where $V_{l,x_i} = span\{\phi_i^{(l)}\}$.

In [5] it was shown that, for a level-wise ordering of the unknowns, the Gauss-Seidel iteration for the semidefinite system corresponds to the multigrid method for the standard system. The switching from grid to grid in the multigrid method corresponds to an outer (block) Gauss-Seidel iteration for the semidefinite system, whereas the MG-smoothing steps resemble an inexact solver for each block by inner Gauss-Seidel iterations. Especially the multigrid V-cycle with one pre- and post-smoothing step by Gauss-Seidel iterations becomes the SGS-method, which is also known as Aitken's double sweep. But other MG-cycle types can be modeled by different orderings of the block GS-traversal. The case of multiple smoothing steps corresponds to multiple inner iterations. Furthermore, other smoothers can be incorporated in the block GS-method as inner iterations.

These algorithms on the semidefinite system (19) can be interpreted alternatively as subspace correction methods [15]. The relaxation of an iterate $u^{(\kappa)} \in V_k$ then takes place with respect to a $\phi \in E_k$ by

$$u^{(\kappa+1)} := u^{(\kappa)} + \lambda \cdot \phi \quad \text{with} \quad a(u^{(\kappa+1)}, \phi) = f(\phi).\tag{25}$$

The different MG-type GS-algorithms for the generating system involve the cyclic application of (25) for a corresponding sequence of functions of E_k.

Note that the proofs of [15] and [18] directly carry over to the Gauss-Seidel iteration for the semidefinite system. There, it is shown that the MG V-cycle that corresponds for our level-wise ordering to SGS has a an error propagation operator and thus a convergence rate independent of k. It behaves like

$$\rho = 1 - 1/C, \quad 1 < C < \infty.\tag{26}$$

If we only iterate the unknowns of the semidefinite system that are associated to the functions of H_k, e.g. if we only use the subset $H_k \subset E_k$ in (25) with some appropriate ordering, then we obtain the hierarchical basis multigrid method of Bank and Yserentant [1]. For practical reasons, the unknowns also have to be ordered level-wise. The error propagation operator of the HB-MG method and thus the convergence rate behave in 2D like

$$\rho = 1 - C/k^2, \quad 1 < C < \infty.\tag{27}$$

If we only iterate the unknowns of the semidefinite system that are associated to the functions contained in B_k, e.g. if we only use the (ordered) subset $B_k \subset E_k$ in (25), we get the standard Gauss-Seidel method for the fine grid system. It's convergence rate behaves like

$$\rho = 1 - C/4^k, \quad 1 < C < \infty. \tag{28}$$

Note that any iteration is convergent if the set of functions involved in one iteration sweep spans V_k. However, (26)-(28) indicate that we have to use E_k itself or at least a sufficiently large subset of E_k to obtain a convergence rate independent of k.

3.2. Point-oriented Block Gauss-Seidel algorithms. Now, we partition the unknowns of the semidefinite system into groups and perform a block-Gauss-Seidel iteration on the associated block-partitioned system. Thus, we decompose the semidefinite matrix L_k^E by $L_k^E = \mathcal{F}_k^E + \mathcal{G}_k^E + (\mathcal{G}_k^E)^T$, where $\mathcal{F}_k^E$ and $\mathcal{G}_k^E$ now denote the block diagonal and strictly lower block triangular parts of L_k^E, respectively. Then, the block GS-iteration is expressed by using $C_k^{E,GS} := (\mathcal{F}_k^E + \mathcal{G}_k^E)^{-1}$, and its symmetric counterpart SGS is expressed by using $C_k^{E,SGS} := (\mathcal{F}_k^E + \mathcal{G}_k^E)^{-T} \mathcal{F}_k^E (\mathcal{F}_k^E + \mathcal{G}_k^E)^{-1}$ in the iteration

$$w_k^{E,(\kappa+1)} := w_k^{E,(\kappa)} + C_k^E (f_k^E - L_k^E w_k^{E,(\kappa)}). \tag{29}$$

Note that $\mathcal{F}_k^E + \mathcal{G}_k^E$ is invertible only for certain choices of partitions, where each group contains unknowns associated to linear independent functions. Nevertheless, if this is not fulfilled, an iterative method still produces a non-unique solution for the corresponding semidefinite subsystem.

Alternatively, one block Gauss-Seidel step can be interpreted as a subspace correction method. The relaxation of an iterate $u^{(\kappa)} \in V_k$ now takes place simultaneously with respect to a set $\Phi \subset E_k$ by

$$u^{(\kappa+1)} := u^{(\kappa)} + \sum_{\phi \in \Phi} \lambda_\phi \cdot \phi \quad \text{with} \quad \forall \phi \in \Phi : a(u^{(\kappa+1)}, \phi) = f(\phi). \tag{30}$$

For the computation of the values of $\lambda_\Phi = \{\lambda_\phi : \phi \in \Phi\}$, the system

$$L_\Phi \lambda_\Phi = r_\Phi \tag{31}$$

with

$$\begin{aligned}
(L_\Phi)_{i,j} &:= a(\phi_j, \phi_i), & \phi_i, \phi_j \in \Phi \\
(r_\Phi)_i &:= (f_\Phi)_i - a(u^{(\kappa)}, \phi_i) \\
(f_\Phi)_i &:= (f, \phi_i), & \phi_i \in \Phi,
\end{aligned} \tag{32}$$

has to be solved. The block Gauss-Seidel method for the semidefinite system is the cyclic application of (30) for a given sequence of (disjoint) subsets of E_k. This type of iteration is only convergent if the union of all involved sets of functions span V_k. Note that the case of non-disjoint subsets can still be denoted in terms of (30), but not more by (29).

Of course, this approach is heavily dependent on the chosen sequence of subsets of E_k. In the following, we suggest a point-oriented approach. This overcomes practical problems that occur with level-oriented multigrid methods if parallelization or adaptive local refinement is considered.

For the parallelization of a MG-algorithm, the domain is usually decomposed into blocks or stripes [12]. The grids covering these subdomains are assigned to different processors. To maintain the functionality of the smoother, data has to be exchanged between the subdomains and thus between the processors. This communication has to take place on all levels and contributes substantially to the overall cost of the parallel algorithm. In the case of adaptive refinement, 'holes' in the grids of certain levels can occur that might hinder efficient smoothing on the respective level.

These problems are overcome by the following approach. We group all unknowns together that are associated to the same grid point. This results in point-oriented methods and can be considered as a point-block technique. Now, an outer iteration switches from grid point to grid point. The local system that belongs to all basis functions of different levels centered in the same grid point can be solved either directly or by an inner iteration running over all unknowns associated to the grid point under consideration. Furthermore, grid points can be grouped together to form subdomains. In this sense, we get some sort of simple domain decomposition method which exhibits MG-type convergence properties.

To be more specific, we consider as blocks the unknowns associated to all functions that are centered in the same grid point $x \in N_k$:

$$P_x := \{\phi \in E_k : \phi(x) = 1\} \tag{33}$$

This corresponds to a point-oriented decomposition

$$V_k = \sum_{x \in N_k} \sum_{l : x \in N_l} V_{l,x} \tag{34}$$

of the space V_k. Note that in comparison with the level-wise decomposition (24) just the summations are exchanged.

Now, we can step through the set N_k of grid points and relax simultaneously the unknowns that belong to the same grid point. This results in systems of linear equations (31) with $\Phi = P_x$, $x \in N_k$, that form the block diagonal matrix $\mathcal{F}_k^E$ involved in (29). Note that the size of the systems belonging to $x \in N_l \setminus N_{l-1}, l = k, .., 2$, is $k - l + 1$. The size of the system belonging to the center point $x = (0.5, 0.5)$ is k. For the case of an operator with constant coefficients, the point-block matrices L_{P_x} are full, definite and symmetric Toeplitz matrices. It turns out that their condition number is $O(1)$. See also Table 3. Thus, the solution of (31) for $x \in N_l \setminus N_{l-1}$ can be obtained by some appropriate iterative method in $O((k - l + 1)^2)$ operations. The number of operations necessary for solving all arising point-block subsystems is in 2D

$$C \frac{3}{4} n_k \left(1^\gamma + \frac{1}{4} 2^\gamma + \frac{1}{16} 3^\gamma + \frac{1}{64} 4^\gamma + ... \frac{1}{4^{k+1-l}} (k + 1 - l)^\gamma + ... + k^\gamma, \right) \tag{35}$$

where $\gamma = 2$ for SGS and $\gamma = 3$ for direct solution by Gaussian elimination. Altogether, this results in $O(n_k)$ operations. The coupling between two point-blocks is described by the respective submatrix entries of $\mathcal{G}_k^E$, or, in (32), by $a(u^\kappa, \phi_i), i = 1, .., |P_x|$, with a non-unique representation (13) of u^κ. It is easy to see that by this coupling information is exchanged on all respective levels of discretization simultaneously.

In practice, however, not all traversal orderings through the set of grid points N_k are advisable. We restrict ourselves to the sequence of point-blocks where the grid points of $N_k \setminus N_{k-1}$ are ordered first, for example in a three-color fashion. Second, the remaining grid points of N_{k-1} are ordered recursively in the same way. Then, neither the point-block matrix nor the off-diagonal blocks have to be assembled explicitly, but can be expressed by means of prolongation and restriction operators. However, the efficient solution of the diagonal-block problems and the block Gauss-Seidel iteration is still tricky to implement. It can be shown that one point-block Gauss-Seidel step can be implemented to need $O(n_k)$ operations. Details are reported in [7].

Note that it is often not necessary to compute the exact solution for each point-block problem. Then, a few GS- or SGS-iterations are sufficient. In the extreme case of one GS step only, we obtain the GS iteration for the semidefinite system (19) with just a special point-oriented ordering.

The point-oriented approach can be easily generalized. We may allow an arbitrary domain decomposition of Ω into K non-overlapping subdomains with associated decomposition

$$N_k = \bigcup_{i=1}^{K} N_k^i, \quad N_k^i \cap N_k^j = \{\} \text{ for } i \neq j \tag{36}$$

of the grid points N_k. Now, we group together the unknowns of the semidefinite system that are associated to functions of E_k whose center points are situated in the same N_k^i by

$$P_{N_k^i} := \bigcup_{x \in N_k^i} P_x. \tag{37}$$

The resulting block Gauss-Seidel algorithm now switches from subdomain to subdomain in some prescribed order. For practical purposes, a nested dissection-like decomposition [4] of the grid points into subdomains is advisable. Compare also Figure 2. Then, by using multigrid prolongation and restriction operators, the submatrices for the subdomains have not to be assembled explicitly, and one overall block Gauss-Seidel iteration can be performed with $O(n_k)$ operations.

Note that in contrast to the point-block approach the arising subproblem matrices are now in general not longer invertible, since they can be semidefinite. However, an iterative method still is able to produce a non-unique solution for the corresponding subsystem. Alternatively, a direct solver for any definite subsystem with full rank can be used.

Now, consider for example the stripe-wise approach in figure 2 (left). All functions of E_k associated to the grid points of different stripes with the same number are

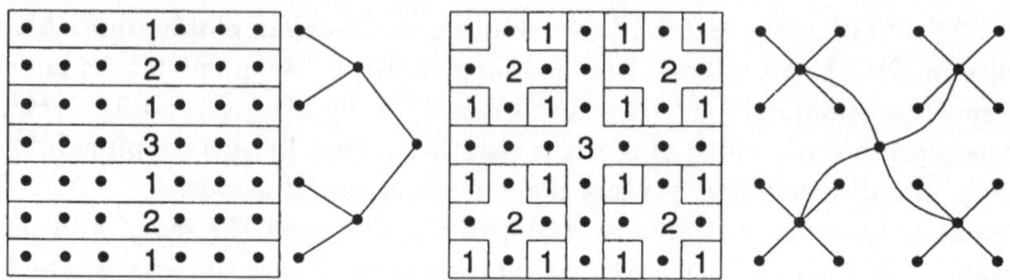

FIG. 2. *Stripe- and box-wise nested dissection decompositions and their parallelization structures.*

mutually orthogonal with respect to $a(.,.)$, since they possess disjoint supports. Therefore, the stripe subproblems with the same numbers can be computed fully in parallel, and we see directly a binary tree ordering for the parallel execution of our algorithm. Communication has to take place only between father and son stripes, i.e. between next-neighbored stripes with successive numbers, and not on each level of discretization like in many parallel MG algorithms.

In contrast to the usual domain decomposition approach, where just the standard basis is used for discretization, the matrices of the subdomain problems are now in general no longer invertible, since they can be semidefinite. For example, the matrix belonging to stripe 3 is semidefinite and has a generalized condition number of $O(1)$. Here, the application of GS automatically results in a MG method. However, the matrices of the subdomain problems of the stripes with number 1 are invertible. Fortunately, they have in general a good condition number independent of k (around 1.666) and GS converges quite fast. For example, for the 2D Laplace operator we obtain here a 1D Laplace operator and a dominating Helmholtz part with forefactor 4^k. We solve these problems either iteratively by a few GS steps or directly by a tridiagonal solver. Similar observations can be made for the problems arising for the intermediate stripes.

For the solution of the subproblems that belongs to one stripe, an inexact solver (one or two point-block SGS-iterations) is also often sufficient. In this sense, the point-block GS method with analogous nested dissection ordering of the stripes (see figure 2) and additional level-wise ordering of the grid points within the stripes is a direct realization of the domain-oriented method with inexact local solver. Thus, the point-block method can be parallelized in the same way as the domain-oriented method. Note that with the special case of each N_k^i containing only one single $x \in N_k$, we obviously have the point-oriented method again.

4. Numerical experiments

We consider the simple model problem

$$\Delta u = f \text{ on } \Omega = (0,1)^2 \tag{38}$$

with Dirichlet boundary conditions.

We measured the reduction rate ρ after a sufficient number κ of iterations as average over the last $\lfloor \kappa/2 \rfloor$ steps by

$$\rho = \left(\frac{\|r_k^\kappa\|_2}{\|r_k^{\lfloor \kappa/2 \rfloor}\|_2} \right)^{1/\lceil \kappa/2 \rceil} \tag{39}$$

with r_k^κ being the residual after κ iterations.

First, we consider the level-wise Gauss-Seidel method for $L_k^E w_k^E = f_k^E$ with such orderings and repetitions that a MG V-cycle with v_1 pre- and v_2 post-smoothing steps is realized. The unknowns on each level are ordered in a four color manner. The results of our experiments are shown in Table 1. Of course, as the level-oriented GS iteration for the semidefinite system is equivalent to MG, these reduction rates are already reported in many papers. We list them here only for comparison with the point-oriented techniques.

TABLE 1

Reduction rates for the level-wise Gauss-Seidel iterating on E_k (MG).

$v1$	$v2$	$k =$	3	4	5	6	7	8
1	0	ρ	0.11	0.11	0.15	0.18	0.20	0.23
0	1	ρ	0.105	0.106	0.107	0.113	0.122	0.127
1	1	ρ	0.020	0.031	0.038	0.043	0.045	0.046
2	1	ρ	0.004	0.006	0.016	0.021	0.023	0.024
1	2	ρ	0.004	0.004	0.014	0.019	0.021	0.022

Now, consider the GS iteration for the semidefinite system that relaxes only the unknowns belonging to the functions of H_k. Thus, we obtain the HB-MG method. Here, we traverse the unknowns in the ordering $(N_k \setminus N_{k-1}, N_{k-1} \setminus N_{k-2}, .., N_1)$. We relax the unknowns of $N_l \setminus N_{l-1}$ v_1 times, where each set $N_l \setminus N_{l-1}$ v_1 is ordered itself in a three color fashion, $l = k, .., 2$. Then, we traverse the unknowns in inversed order and relax each set $N_l \setminus N_{l-1}$ v_2 times. The observed reduction rates are shown in Table 2. The dependency (27) of the reduction rates from the number k of levels can be seen clearly.

TABLE 2

Reduction rates for the level-wise Gauss-Seidel iterating on H_k (HB-MG).

$v1$	$v2$	$k =$	3	4	5	6	7	8
1	0	ρ	0.549	0.660	0.737	0.794	0.833	0.863
0	1	ρ	0.537	0.648	0.729	0.787	0.827	0.858
1	1	ρ	0.445	0.571	0.659	0.724	0.773	0.810
2	1	ρ	0.424	0.553	0.642	0.710	0.758	0.800
1	2	ρ	0.409	0.538	0.633	0.696	0.749	0.788

The point-block method relaxes simultaneously *all* unknowns that belong to the functions of E_k centered at the same grid point, whereas in the HB-MG method only the unknown is relaxed that belongs to the function with the largest support. In this sense, the point-block method can be seen as an extension of the hierarchical basis MG method that overcomes its disadvantage with respect to the convergence behavior. In contrast to standard MG, all algorithmical advantages of HB-MG remain, specially with respect to parallelization and adaptivity.

In Table 3, we show the condition numbers of the matrix (32) belonging to one single grid point (here the point (0.5,0.5)) for different k. Furthermore, the spectral radius of the symmetric Gauss-Seidel iteration relaxing on the point-block system is shown. We see that the point-block subsystems can be solved iteratively by a number

TABLE 3

Condition number and SGS spectral radius for the submatrix belonging to one point-block.

$k =$	5	10	15	20	30	40	50	100	200	300
κ	11.735	15.612	16.873	17.429	17.889	18.070	18.161	18.287	18.321	18.326
ρ	0.4365	0.4426	0.4436	0.4440	0.4442	0.4443	0.4444	0.4444	0.4444	0.4444

of SGS-steps independent of k. SGS converges at least with a convergence factor of 0.4444. A similar observation holds for the spectral radius and the convergence factor of GS.

The reduction rate for the point-block GS method is shown in Table 4. We see that it is independent of k. Here, the ordering of the grid points is the same as in the HB-MG experiments described above. Note that for $v_1 = 2$ or $v_2 = 2$ no substantial improvement of the rates can be observed. With respect to efficiency, the best cases are $(v_1, v_2) = (0, 1)$ and $(v_1, v_2) = (1, 0)$, representing just a point-block GS sweep without any inner repetition of certain grid points.

TABLE 4

Reduction rates for the point-block Gauss-Seidel, exact solution of the point-block systems.

$v1$	$v2$	$k =$	3	4	5	6	7	8
1	0	ρ	0.132	0.136	0.137	0.145	0.154	0.156
0	1	ρ	0.125	0.133	0.135	0.136	0.139	0.140
1	1	ρ	0.035	0.045	0.051	0.055	0.057	0.058
2	1	ρ	0.032	0.043	0.051	0.054	0.056	0.057
1	2	ρ	0.033	0.055	0.062	0.063	0.065	0.066

If we substitute the exact solver for the arising point-block subproblems by one SGS-iteration, we obtain the reduction rates shown in Table 5. Using 1 GS step only, the point-block GS method deteriorates to a simple GS iteration for the semidefinite system (19) with a special point-oriented ordering. The amount of work is comparable to the level-oriented GS algorithm. Note that there is no substantial difference to the rates of the first and second row of Table 1. Thus, its efficiency is comparable to the respective MG-method. However, in contrast to the point-block method with exact inner solver, we see no improvement of the reduction factors for the case $v_1 = 1, v_2 = 1$.

TABLE 5

Reduction rates for the point-block Gauss-Seidel, inexact solution of the point-block systems by one SGS-step.

$v1$	$v2$	$k =$	3	4	5	6	7	8
1	0	ρ	0.210	0.245	0.265	0.284	0.294	0.299
0	1	ρ	0.133	0.194	0.234	0.263	0.279	0.288
1	1	ρ	0.173	0.218	0.248	0.271	0.288	0.301

Now, consider the domain-oriented block GS iteration with stripe-like decomposition and nested dissection ordering like in Figure 2. The observed reduction rates are shown in Table 6. The obtained rates seem to be independent of k. However, in comparison with the results of Table 4, no improvement but even a deterioration of the reduction rates can be observed. Thus, in comparison with the point-oriented block GS method, the exact solution of the subproblems on the stripes is not necessary and less efficient. It can be substituted by one or two point-block GS steps working on the stripe subdomains. This would result in a point-block GS method with a slightly different ordering as above.

TABLE 6

Reduction rates for the stripe-wise domain decomposition block Gauss-Seidel.

$v1$	$v2$	$k =$	3	4	5	6	7	8
1	0	ρ	0.076	0.145	0.188	0.210	0.225	0.229
0	1	ρ	0.077	0.152	0.184	0.203	0.216	0.223
1	1	ρ	0.053	0.113	0.140	0.164	0.174	0.179

We expect the domain decomposition of Figure 2 (right) to perform better with respect to reduction rate and efficiency. The arising subdomain problems can be solved by a few SGS-iterations since they are, due to the generating system, automatically equivalent to a MG-like method. Respective experiments will be carried out soon.

5. Concluding remarks

In this paper we presented different multilevel algorithms based on the generating system approach. We studied level-oriented techniques where GS methods for the arising semidefinite system turn out to result either in conventional MG algorithms or in the HB-MG method. Additionally, we presented new point- (and domain-) oriented methods. There, block GS iterations for the semidefinite system are obtained that exhibit a reduction rate independent of the grid size and number of levels like conventional MG methods. Furthermore, these algorithms possess favorable properties with respect to parallelization and adaptivity. In contrast to the parallelization of a MG method, where communication has to take place on every level of discretization, our new algorithms have communication requirements like conventional domain decomposition methods. Thus, we demonstrated that the use of the generating system is an excellent guideline for the development of new multilevel algorithms.

The theory for proving k-independent convergence rates for our algorithms has to follow the lines of Xu [15] and Zhang [18]. For the level-oriented approach, these proves can be rewritten directly in terms of the generating system. For the point-oriented approach, this will be done in near future. Anyway, we believe that for any ordering of the semidefinite system a convergence behavior like $1 - 1/C, 1 < C < \infty$, can be shown. Thus, it is not more necessary to maintain a level-wise ordering, but other orderings might be found that can be implemented to need $O(n_k)$ operations per iteration step and exhibit advantages for parallelization and adaptivity.

Note that the point- and domain-oriented approach is closely related to the frequency filtering ILU technique [14], but shows a stronger grid oriented flavor.

45

In [6], our level- and point-oriented approach is adopted to an extended generating system that contains additionally the basis functions of all grids that result from semi-refinement steps with respect to both coordinate directions. Furthermore, the full and the sparse grid case [17] is considered. For the level-oriented methods, we obtain MG algorithms similar to that of Hackbusch [8], [9], Naik and van Rosendale [11], Mulder [10] and Wesseling [13]. For the point-oriented approach analogous point-block Gauss-Seidel methods as well as BPX-like preconditioners are derived.

REFERENCES

[1] R. BANK, T. DUPONT, AND H. YSERENTANT, *The hierarchical basis multigrid method*, Num. Math., 52 (1988), pp. 427–458.

[2] J. BRAMBLE, J. PASCIAK, AND J. XU, *Parallel multilevel preconditioners*, Math. Comp., 31 (1990), pp. 333–390.

[3] H.-J. BUNGARTZ, *Dünne Gitter und deren Anwendung bei der Lösung der dreidimensionalen Poisson-Gleichung.*, Promotion.

[4] A. GEORGE, *Nested dissection of a regular finite element mesh*, SIAM, J. Numer. Anal. 10, 1973.

[5] M. GRIEBEL, *Multilevel algorithms considered as iterative methods on indefinite systems*, TU München, Institut f. Informatik, TUM-I9143; SFB-Report 342/29/91 A, 1991.

[6] M. GRIEBEL, C. ZENGER, AND S. ZIMMER, *Improved multilevel algorithms for full and sparse grid problems*, TU München, Institut f. Informatik, SFB-Report, 1992.

[7] M. GRIEBEL AND S. ZIMMER, *On practical aspects of multi-level point block relaxation schemes*, TU München, Institut f. Informatik, SFB-Report, 1992.

[8] W. HACKBUSCH, *A new approach to robust multi-grid solvers*, Bericht, Institut für Informatik und praktische Mathematik, CAU, Kiel, 1987.

[9] ———, *The frequency decomposition multi-grid method, part I: applications to anisotropic equations*, Numerische Mathematik, Vol.—, 1989.

[10] W. MULDER, *A new multigrid approach to convection problems*, CAM Report 88-04, (1988).

[11] N. H. NAIK AND J. VAN ROSENDALE, *The improved robustness of multigrid elliptic solvers based on multiple semicoarsened grids*, ICASE report 51-70, (1991).

[12] O. MCBRYAN ET AL., *Multigrid methods on parallel computers - a survey of recent developments*, Impact of Computing in Science and Engineering 3, pp. 1-75, 1991.

[13] P. WESSELING, *The method of Mulder, Naik and van Rosendale*, talk given in Oberwolfach, 1992.

[14] G. WITTUM, *Filternde Zerlegungen, schnelle Löser für große Gleichungssysteme*, Teubner Scripten zur Numerik, Hrsg. H. Bock, W. Hackbusch, R. Rannacher, 1992.

[15] J. XU, *Iterative Methods by Space Decomposition and Subspace Correction: A Unifying Approach*, Report No. AM67, Department of Mathematics, Penn State University, revised 1992.

[16] H. YSERENTANT, *On the multi-level splitting of finite element spaces*, Numer. Math., 49 (1986), pp. 379–412.

[17] C. ZENGER, *Sparse Grids*, in Parallel Algorithms for Partial Differential Equations, Proceedings of the Sixth GAMM-Seminar, Kiel, January 19-21, 1990, W. Hackbusch, ed., Vieweg-Verlag, 1991.

[18] X. ZHANG, *Multilevel Schwarz methods*, University of Maryland, Department of Computer Science, Tech. Report CS-TR-2894, UMIACS-TR-92-52, 1992.

PARALLELIZABLE INCOMPLETE FACTORIZATION PRECONDITIONING METHODS

IVAR GUSTAFSSON and GUNHILD LINDSKOG
Department of Computer Sciences, Chalmers University of Technology
S-412 96 Göteborg, Sweden

Summary

Two classes of parallelizable preconditioning iteration methods for the solution of finite difference or finite element matrix problems are presented. The first class is developed with regard to massively parallel computers and we discuss various methods to construct completely parallel preconditioners. The second class is intended for computers with a smaller number of parallel processors and uses a blockwise parallel incomplete factorization. Both types of methods are based on the calculation of approximate inverses. The speed of the methods may be increased by making the approximation of the inverses more accurate. Problems with constant as well as strongly varying orthotropy are examined and the methods are compared with respect to the computational complexity.

1. Introduction

We consider finite difference or finite element discretizations of the second order orthotropic boundary value problem,

$$-\frac{\partial}{\partial x}(a_1(x,y)\frac{\partial u}{\partial x}) - \frac{\partial}{\partial y}(a_2(x,y)\frac{\partial u}{\partial y}) = f(x,y)$$

on a rectangular domain Ω. The functions $a_i(x,y), i = 1, 2$ are positive and bounded. We assume a Dirichlet condition on the boundary of the domain.

The resulting system of linear equations $Ax = b$, where A is a symmetric positive definite matrix, is solved by the preconditioned conjugate gradient method, PCG, [1].

The purpose of the paper is to present various possibilities for performing the PCG method in parallel. Two degrees of parallelity are considered, completely and blockwise parallel preconditioners. The degree of parallelity is valid both for the calculation of the preconditioner and the solution of the preconditioning system.

Both classes of methods are mainly based on incomplete factorization. In the first class , the preconditioning system is solved by a matrix vector multiplication with $\hat{A}$, where $\hat{A}$ is an approximate inverse of A, i.e. an explicit calculation process. The preconditioning matrix, called C, is defined by $C = \hat{A}^{-1}$. We present various methods for the approximation of inverses. The complete parailelity makes the methods suitable for use on massively parallel computers.

The second class of methods is based on the relaxed block incomplete factorization presented in [3]. In the present paper we give a blockwise parallel version of the method. The parallelity is obtained by approximating the inverses of the successively calculated Schur complements. The solution of the preconditioning system consists of a backward and a forward solution step on blocklevel, i.e. an implicit calculation process. On each blocklevel the linear system is solved by explicit calculations. The methods are preferably used on computers with a smaller number of parallel processors. These methods are applicable to more general problems than the methods in the previous papers [2], [4].

The completely and the blockwise parallel methods are presented in Sections 2 and 3 respectively. In Section 4 we compare the computational complexity for all presented methods and examine problems with constant and strongly varying orthotropy with respect to the rate of convergence and the amount of computational work.

2. Completely parallel preconditioners

The first suggestion for a completely parallel preconditioner is based on the SSOR factorization of A, [1],

$$A_{SSOR} = (\frac{D}{\omega} + E)(\frac{D}{\omega})^{-1}(\frac{D}{\omega} + E^T), \tag{2.1}$$

where $A = D + E + E^T$, $D = diag(A)$ and E is the strictly lower triangular part of A. The matrix (2.1) can also be written

$$A_{SSOR} = (I + L)\tilde{D}(I + L^T), \tag{2.2}$$

where $L = \omega E D^{-1}$, $\tilde{D} = \frac{1}{\omega}D$ and I is the identity matrix. The use of (2.2) as a preconditioner defines a sequential calculation process in the solution of the preconditioning system. Hence, instead of using (2.2) we calculate an approximate inverse of A_{SSOR} giving the solution by a completely parallel matrix vector multiplication. The exact inverse of A_{SSOR} is given by

$$A_{SSOR}^{-1} = (I + L^T)^{-1}\tilde{D}^{-1}(I + L)^{-1}.$$

An approximate inverse of A_{SSOR} and hence of A is calculated from an approximate inverse $\hat{L}$ of $I + L$.

For $\| L \| < 1$ we have

$$(I + L)^{-1} = \sum_{k=0}^{\infty} (-1)^k L^k .$$

Let

$$\hat{L} = \sum_{k=0}^{p} (-1)^k L^k \ , p \geq 1.$$

As an approximate inverse of A we may now define

$$\hat{A} = \hat{L}^T \tilde{D}^{-1} \hat{L} . \tag{2.3}$$

Since

$$(\hat{A}x, x) = (\hat{L}^T \tilde{D}^{-1} \hat{L}x, x) = (\tilde{D}^{-\frac{1}{2}}\hat{L}x, \tilde{D}^{-\frac{1}{2}}\hat{L}x) > 0 , \forall x \neq 0,$$

the matrix $\hat{A}$ is positive definite and the preconditioning matrix $C = \hat{A}^{-1}$ exists. Obviously the matrix C is not calculated.

Since we do not need $\hat{A}$ explicitely we just use information from L and $\tilde{D}$ and these matrices are calculated in parallel over the total number of unknowns. The solution of the preconditioning system is equivalent to the calculation of $\hat{L}^T \tilde{D}^{-1} \hat{L}$ times a vector. For $p > 1$ the calculation of $\hat{L}$ times a vector is done by successive matrix vector multiplications by L and similarly for $\hat{L}^T$. These operations are also possible to do in parallel over the total number of unknowns.

In a forthcoming paper, presently under preparation, we show that for $0 < \omega < 2$, independent of h, the condition number $\kappa (C^{-\frac{1}{2}} A_{SSOR} C^{-\frac{1}{2}})$ is of order $O(1)$, $h \rightarrow 0$. For the model problem and $\omega = 2 - \xi h$, where ξ is a parameter independent of h, this condition number is of order $O(h^{-1})$, $h \rightarrow 0$. It is well known, see eg [1], that the SSOR method gives a condition number for the preconditioned system of order $O(h^{-2})$ for ω independent of h and of order $O(h^{-1})$ for $\omega = 2 - \xi h$. Hence, in all cases, the total condition number is of order $O(h^{-2})$, $h \rightarrow 0$.

We also present two alternative possibilities for the choice of the approximate inverse $\hat{A}$ of A. In the first we choose a symmetric set of indices S, where nonzero entries are allowed in $\hat{A}$. $\hat{A}$ is then calculated from the relation

$$(\hat{A}A)_{k,j} = I_{k,j} \ , (k , j) \in S. \tag{2.4}$$

If for instance, $\hat{A}$ has a five diagonal structure, a 5×5 positive definite linear system has to be solved for each row of $\hat{A}$. Since this implies an unsymmetric matrix $\hat{A}$, symmetrization has to be done before using it as a preconditioner. Since in this method, $\hat{A}$ may not be positive definite, we also present a second alternative where this property is valid. To this end we choose $\hat{A}$ as a product of triangular matrices,

$$\hat{A} = LL^T , \tag{2.5}$$

where $L = \hat{L}\tilde{D}$ and $\hat{L}$ is a lower triangular approximate inverse of A such that

$$(A\hat{L})_{k,j} = I_{k,j} \ , (k,j) \in S$$

and S is a set of indices where nonzero entries are allowed in $\hat{L}$. $\tilde{D}$ is a diagonal matrix, $\tilde{D} = diag\,(\hat{L})^{-\frac{1}{2}}$, making $diag\,(\hat{A}A) = I$. The matrix $\hat{A}$ in (2.5) is positive definite, since $diag\,(L) > 0$. Hence the preconditioning matrix $C = \hat{A}^{-1}$ exists.

3. A blockwise parallel preconditioner

The complete block factorization of a symmetric tridiagonal matrix

$$A = \begin{bmatrix} A_{11} & A_{12} & & & \\ A_{21} & & & & \\ & & \ddots & & \\ & & & & A_{N-1,N} \\ & & & A_{N,N-1} & A_{NN} \end{bmatrix} \tag{3.1}$$

is given by

$$\begin{bmatrix} X_1 & & & \\ A_{21} & X_2 & & \\ & & \ddots & \\ & & A_{N,N-1} & X_N \end{bmatrix} \begin{bmatrix} X_1^{-1} & & & \\ & X_2^{-1} & & \\ & & \ddots & \\ & & & X_N^{-1} \end{bmatrix} \begin{bmatrix} X_1 & A_{12} & & \\ & X_2 & A_{23} & \\ & & \ddots & A_{N-1,N} \\ & & & X_N \end{bmatrix} \tag{3.2}$$

where

$$X_1 = A_{11}, \ X_i = A_{ii} - A_{i,i-1} X_{i-1}^{-1} A_{i-1,1}, i = 2, \ldots, N . \tag{3.3}$$

Our aim is to choose a preconditioner based on the factorization (3.2). Hence, the latter is approximated in order to get an easily solvable preconditioning system.

In the relaxed incomplete block factorization studied in [3], the matrices are defined by

$$X_1 = A_{11}, X_i = A_{ii} - A_{i,i-1} \hat{X}_{i-1}^{(p)} A_{i-1,i} - \omega \Delta_{i-1}, i = 2, \ldots, N , \tag{3.4}$$

where $\hat{X}_{i-1}^{(p)}$ is an approximation of X_{i-1}^{-1} with bandwidth p located symmetrically about the main diagonal and calculated as in (2.4), ω is a relaxation parameter and Δ_{i-1} is a diagonal matrix introduced by the reason as follows.

With the block partitioning of A given in (3.1), let $A = D - L - L^T$, where D and $-L$ are the respectively block diagonal and strictly lower block triangular matrices of A. The matrix X is block diagonal with blocks made up of X_i , $i = 1, \ldots, N$. The preconditioning matrix can then be written

$$C = (X - L) X^{-1} (X - L^T) = X + LX^{-1}L^T - L - L^T = A + R$$

where $R = X - D + LX^{-1}L^T$ is the error matrix. Analogous to the pointwise incomplete factorization, [5], a rowsum criterion

$$Re = 0 , e^T = (1, 1 , \ldots , 1) , \omega = 1 \tag{3.5}$$

is introduced. Computer experiments have shown that such a rowsum criterion implies a faster convergence for certain model problems also for block methods. The entries of the matrix Δ_{i-1} are calculated such that the condition (3.5) is fulfilled. By the relaxation parameter ω in (3.4) the rate of convergence may be improved at least for model problems.

The solution of the preconditioning system with the choice of X_i in (3.4) implies a sequential calculation process both concerning the dependence between the bock levels and within a block level.

A method which is parallel within each block level is obtained by using the matrices $\hat{X}_{i-1}^{(p)}$ not only in the calculation of the matrices X_i but also in the solution of the block-diagonal systems.

Consider the lower block triangular part of the preconditioning system, i.e. $(X - L)h = g$. We assume a block partioning of h and g according to (3.1). Instead of solving

$$X_1 h_1 = g_1 , X_i h_i = g_i - A_{i,i-1} h_{i-1} , i = 2 , \ldots , N$$

the vectors h_i are calculated by

$$h_1 = \hat{X}_1^{(p)} g_1 , h_i = \hat{X}_i^{(p)} (g_i - A_{i,i-1} h_{i-1}) , i = 2 , \ldots , N. \tag{3.6}$$

Writing the preconditioning system on the form

$$C = (X - L) X^{-1} (X - L^T) = (X - L) (I - X^{-1}L^T) \tag{3.7}$$

also the upper block triangular part is solved by multiplications by $\hat{X}_i^{(p)}$, $i = 1 , \ldots , N-1$. Hence, all calculations within a block may be done in parallel.

Since the diagonal blocks of X in the preconditioning matrix C now are made up of the true inverses of the approximate inverses of X_i we have to change the notation X in C, see (3.7), in order to be able to express the matrix C. We use the notation $\overline{X}$.

C is now defined by

$$\overline{X}_1 = (\hat{X}_1)^{-1} \, , \overline{X}_i = (\hat{X}_i)^{-1} \, , i = 2 , \ldots , N \, , \tag{3.8}$$
$$\hat{X}_1 = \tilde{X}_1^{(p)} - \omega \, \Delta_0 \, , \hat{X}_i = \tilde{X}_i^{(p)} - \omega \, \Delta_{i-1} \, ,$$

and $\tilde{X}_i^{(p)}$ is a bandmatrix approximating $X_i^{-1} \, , i = 1 , \ldots , N$ and

$$X_1 = A_{11} \, , X_i = A_{ii} - A_{i,i-1} \, \hat{X}_{i-1}^{(p)} \, A_{i-1,i} \, , i = 2 , \ldots , N \, .$$

We observe that the matrices $X_i \, , i = 1 , \ldots , N$ never need to be calculated. The relaxation parameter ω and the diagonal matrices $\Delta_{i-1} \, , i = 1 , \ldots , N$ are introduced in order to satisfy the rowsum criterion (3.5). In the numerical tests p is chosen as 3 or 5.

4. Comparisons and conclusions

The following preconditioners are compared, where (1) - (5) are the completely parallel methods:

   (1) No preconditioning
   (2) The diagonal of A
   (3) Approximate inverses of the SSOR factorization, 2 - 5 terms, see (2.3)
   (4) An ansatz of a five diagonal matrix, see (2.4)
   (5) A product of triangular matrices, see (2.5)
   (6) The block incomplete factorization with 3 or 5 diagonals, see (3.8) .

The computational work for the calculation of the preconditioner and the work per iteration in the preconditioned conjugate gradient method is given in Table 4.1. We assume a 5-diagonal structure of the matrix A. In one flop we include one multiplication and one addition.

Table 4.1   The computational work given as the number of flops per unknown for (A)-
the preconditioner, (B)-the solution of the preconditioning system and (C)-
the remaining calculations in one PCG iteration.

	(1)	(2)	(3) Number of terms				(4)	(5)	(6) Number of diagonals	
			2	3	4	5			3	5
(A)	-	-	2	2	2	2	9	9	11	42
(B)	-	1	5	9	13	17	5	6	8	12
(C)	11	11	11	11	11	11	11	11	11	11

First we consider a problem with constant orthotropy, (if $\varepsilon \neq 1$)

$$-\varepsilon u_{xx} - u_{yy} = 1 \,, (x,y) \in \Omega = (0,1) \times (0,1) \tag{4.1}$$

$$u = 0 \,,(x,y) \in \delta\Omega .$$

The problem is discretized by the five-point difference approximation on a rightan-
gled triangulation of Ω and a rowwise numbering of the meshnodes. In the case $\varepsilon = 100$
a columnwise nodeordering is used in the blockwise method in order to get the same
favourable matrix as in the case $\varepsilon = 0.01$ with small bidiagonal elements compared to
the main diagonal, which implies a high accuracy in the approximate inverses.

The number of iterations and the number of flops per unknown for the relative stopping criterion

$$\| r^k \|_2 \le 10^{-4} \| r^0 \|_2 \tag{4.2}$$

and $\varepsilon = 0.01$, 1 and 100 are given in Table 4.2, Table 4.3 and Figure 4.1.

Table 4.2 The number of iterations for various preconditioners and problem (4.1),
$\varepsilon = 0.01, 100$

h^{-1}	(1)	(3) Number of terms				(6) Number of diagonals	
		2	3	4	5	3	5
8	11	9	8	7	6	1 ($\omega = 1$)	1 ($\omega = 1$)
16	32	20	15	13	11	2 ($\omega = 1$)	2 ($\omega = 1$)
32	83	46	34	28	24	3 ($\omega = 1$)	2 ($\omega = 1$)
64	194	101	72	58	49	5 ($\omega = 1$)	4 ($\omega = 1$)

Table 4.3 The number of iterations for various preconditioners and problem (4.1), $\varepsilon = 1$.

h^{-1}	(1)	(3) Number of terms				(4)	(5)	(6) Number of diagonals	
		2	3	4	5			3	5
8	9	7 ($\omega = 1$)	6	6	6	6	8	5 ($\omega = 0.2$)	4 ($\omega = 0.2$)
16	20	12 ($\omega = 0.7$)	10	8	8	11	13	9 ($\omega = 0.4$)	6 ($\omega = 0.4$)
32	41	24 ($\omega = 0.7$)	17	13	13	21	25	15 ($\omega = 0.6$)	10 ($\omega = 0.6$)
64	84	43 ($\omega = 0.6$)	30	24	21	42	45	25 ($\omega = 0.8$)	15 ($\omega = 0.8$)

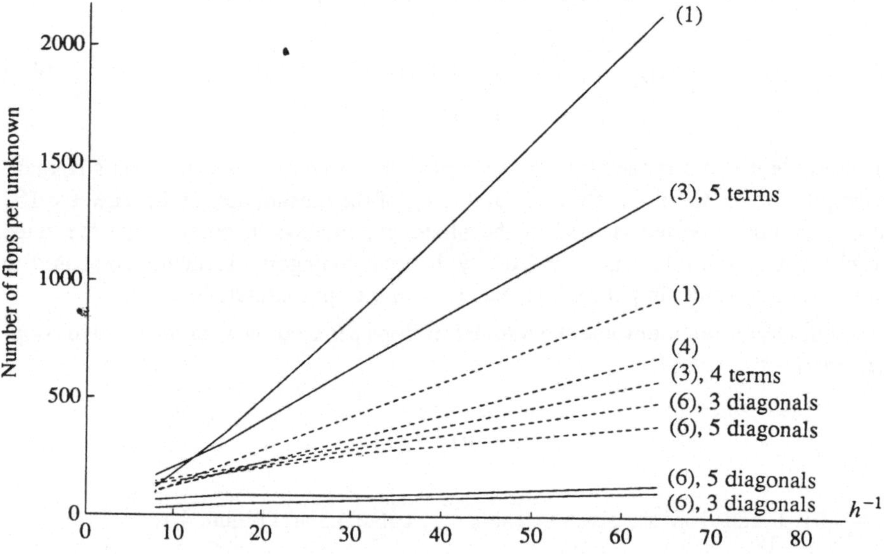

Figure 4.1 The number of flops per unknown. Bold lines: $\varepsilon = 0.01, 100$, dashed lines: $\varepsilon = 1$.

Next, the following problem with strongly varying orthotropy is examined,

$$-\frac{\partial}{\partial x}((0.01 + 100\, x^2 y^2)\, \frac{\partial u}{\partial x}) - \frac{\partial^2 u}{\partial y^2} = 1, (x, y) \in \Omega = (0,1) \times (0,1) \qquad (4.3)$$

$$u = 0, (x, y) \in \delta\Omega.$$

Table 4.4 and Figure 4.2 show the number of iterations and the computational work for the stopping criterion (4.2).

Table 4.4 The number of iterations for problem (4.3)

h^{-1}	(1)	(2)	(3) Number of terms				(4)	(5)	(6) Number of diagonals	
			2	3	4	5			3	5
8	40	22	12 ($\omega = 0.7$)	9	8	7	13	11	6 ($\omega = 0$)	4 ($\omega = 0$)
16	139	45	22 ($\omega = 0.7$)	17	14	13	24	24	12 ($\omega = 0$)	8 ($\omega = 0$)
32	387	93	47 ($\omega = 0.7$)	34	28	25	51	50	24 ($\omega = 0$)	16 ($\omega = 0$)
64	941	190	97 ($\omega = 0.6$)	69	57	49	104	105	51 ($\omega = 0$)	32 ($\omega = 0$)

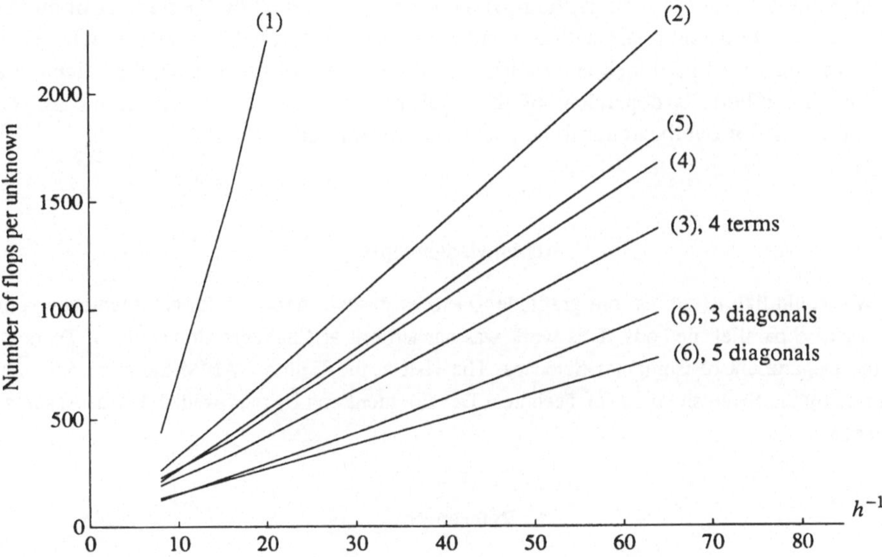

Figure 4.2 The total number of flops per unknown in the solution of problem (4.3) and the stopping criterion (4.2).

It is observed that the number of iterations as a function of h is well in agreement with the theoretical result $O(h^{-1})$, $h \to 0$ for the considered preconditioned CG methods. As the number of terms is increased in the method (3), the rate of convergence approaches the true SSOR method. In case of large orthotropy, the rate of convergence of the SSOR method is $O(h^{-1})$. For the problem (4.1), $\varepsilon = 1$ however, the number of iterations for the true SSOR method is of order $O(h^{-\frac{1}{2}})$, $h \to 0$. The optimal value of ω is given in some cases.

As is already explained, the block incomplete factorization preconditioner (6) implies a rapid convergence for the problem (4.1) and $\varepsilon = 0.01$ and 100, while for $\varepsilon = 1$

the number of iterations behaves like $h^{-0.7}$ for the considered values of h. For problem (4.3) the observed behaviour is like h^{-1} .

Taking the computational complexity and the number of iterations into account, the best choice of the completely parallel methods will be the SSOR based method with 4-5 terms as is seen from the Figures 4.1 and 4.2. A greater number of terms in this method is shown to give a totally larger amount of work. The best choice of the SSOR based method decreases the number of flops per unknown by about 36% for problem (4.1) and by 40% for problem (4.3) compared to the unpreconditioned CG method in case (4.1) and the diagonal preconditioner in case (4.3). For the blockwise parallel method and problem (4.1), $\varepsilon = 1$ and problem (4.2), the corresponding decrease is about 58% and 65% respectively.

Implemented on a parallel computer, where the number of parallel processors exeeds the number of unknowns, the time for the calculations in the completely parallel methods (1)-(4), apart from overhead, theoretically can be estimated from the number of flops in Figures 4.1 and 4.2. In method (6) the time is dependent on the number of blocks in the matrix. On a computer with a smaller number of parallel processors, where the number of unknowns per block is a multipel of the number of processors, the calculation time for all methods is dependent of this multipel and on the number of blocks. Of course, the time for overhead has to be taken into consideration in all cases.

Acknowledgements

We would like to express our gratitude to Professor Owe Axelsson for communications on the blockwise parallel methods. This work was carried out at Chalmers University of Technology, the Department of Computer Sciences. The visit at the Eighth GAMM-Seminar, Kiel was supported by the Swedish board for Technical Development and by the Swedish Royal Academy of Sciences.

References

[1]   O. AXELSSON and V. A. BARKER, *Finite element solution of boundary value problems. Theory and computation.* Academic Press, New York, 1984

[2]   O. AXELSSON, G. CAREY and G. LINDSKOG, *On a class of preconditioned iterative methods on parallel computers,* Int. J. Numer. Meth. Eng., Vol. 27, 637-654 (1989)

[3]   O. AXELSSON and G. LINDSKOG, *On the eigenvalue distribution of a class of preconditioning methods,* Numer. Math. 48, 479 - 498 (1986)

[4]   O. AXELSSON and G. LINDSKOG, *Constant wavefront iteration methods for nine- and 15-point difference matrices,* Computing 46, 233-252 (1991)

[5]   I. GUSTAFSSON, *Modified incomplete Cholesky (MIC) methods,* in: D. J. Evans, ed., Preconditioning methods; Theory and applications, (Gordon and Breach, New York, 1983) 265-293

Modification of the ILU-Method
for Enhanced Parallel Efficiency

Graham Horton[1], Ralf Knirsch[1] and Gabriel Wittum[2]

[1] Lehrstuhl für Rechnerstrukturen (IMMD3)
Universität Erlangen-Nürnberg
Martensstr. 3, D-8520 Erlangen

[2]Institut für Angewandte Mathematik
Universität Heidelberg
Im Neuenheimer Feld 294, D-6900 Heidelberg

Abstract

We consider the ILU method as an iteration for the solution of a two-dimensional p.d.e.
when discretized on a regular square grid. It is known that a parallelization of this method
is possible when the dimension of the processor array is less than that of the problem.
However, in terms of parallelism, ILU methods suffer from two disadvantages: the procedure is not well vectorizable and it has a very fine granularity, making it unsuitable for
some parallel architectures. In this presentation, two modifications to the method will
be used to improve this state of affairs. A blocking method, which lumps communications together is used to coarsen the granularity. A modification to the computation of
the which essentially removes the recursion is shown to enable both vectorization and a
coarser granularity of parallelism. Experimental results demonstrate both the numerical
behaviour of the schemes and the improvements in efficiency obtained on two DMMP
parallel computers.

1. Introduction

We consider the numerical solution of partial differential equations with iterative solvers
based on incomplete decompositions. This solution method may be the incomplete decomposition method itself, a conjugate-gradient method with ILU preconditioning, or a
multigrid method with ILU used as a smoother. Our experimental results will be drawn
from the latter case. For simplicity we shall consider the two-dimensional Poisson problem
discretized on the unit square by a regular, rectangular grid using the five-point stencil.

ILU methods may be parallelized on an MIMD ($\underline{M}$ultiple $\underline{I}$nstruction $\underline{M}$ultiple $\underline{D}$ata)
multiprocessor in a pipeline fashion when the dimension of the processor array is less
than that of the problem, i.e. a 2-D problem may be parallelized on a linear array of
processors. The parallel solution procedure is characterized by a very fine granularity,
will be explained in section 2. This means that the method will not run efficiently on
those parallel machines with a high message latency. In the interest of parallel efficiency,

therefore, a coarser granularity, i.e. a smaller number of messages per iteration, is to be preferred. One way of achieving this without a modification of the computation is to block the messages and the corresponding computations, as has been suggested by Bastian and Horton in [1], and more recently by Manneback and Qin in [4]. This approach is described in section 3.

The second class of parallel architectures according the well-known taxonomy of Flynn [2] is that of the SIMD (_S_ingle _I_nstruction _M_ultiple _D_ata) machines. This class includes both vector and array processors. A parallelization of the ILU decomposition on this type of machine is prevented by the recursive nature of the computation. In section 3 we present a modification of the ILU decomposition which contains no recursion, and which is thus ideally suited to run on SIMD computers. One additional advantage of this approach when applied to MIMD machines is that the granularity of the decomposition achieves a maximum coarseness (i.e. a minimal number of messages).

In section 4 we examine the properties of the modified decomposition; in particular we establish the smoothing property for the model problem. Numerical experiments were performed on two different parallel computers - a Transputer system and a SUPRENUM cluster. The latter machine is particularly interesting, since it has both a high message latency and also a vector processing ability and should benefit doubly from the algorithmic modifications proposed in this paper. The results of the experiments are presented in section 5.

2. The Standard ILU-Factorization

As a model problem we used the Poisson equation

$$-\Delta u = f$$

in the unit square. The discretization with a standard five-point stencil and lexicographical ordering of the unknowns leads to a pentadiagonal system matrix A. We denote the non-zero diagonals of A by A^N, A^S, A^W, A^E and A^C (Center), according to the position of the corresponding coefficient in the discretization molecule. The zero-order incomplete LU-factorization is defined as

$$A = LU - C$$

where L and U denote a lower and upper triangular matrix, respectively and C is the remainder matrix. In addition, we require $LU = A$ on $patt(A)$, where $patt(A)$ denotes the pattern of nonzero entries of the matrix A. Furthermore L and U must satisfy $patt(L) \cup patt(U) = patt(A)$. The computation of the entries of the matrices L and U is performed as follows:

$$L_{i,j}^S = A_{i,j}^S \tag{1}$$
$$L_{i,j}^W = A_{i,j}^W \tag{2}$$

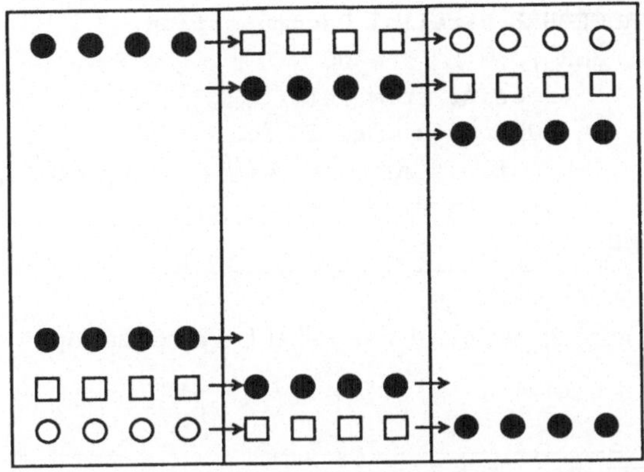

Figure 1: Parallelization of the ILU method

$$L_{i,j}^{C} = A_{i,j}^{C} - \frac{A_{i,j}^{W} * A_{i-1,j}^{E}}{L_{i-1,j}^{C}} - \frac{A_{i,j}^{S} * A_{i,j-1}^{N}}{L_{i,j-1}^{C}} \tag{3}$$

$$U_{i,j}^{E} = \frac{A_{i,j}^{E}}{L_{i,j}^{C}} \tag{4}$$

$$U_{i,j}^{C} = 1 \tag{5}$$

$$U_{i,j}^{N} = \frac{A_{i,j}^{N}}{L_{i,j}^{C}} \tag{6}$$

$$C_{i,j}^{SE} = \frac{A_{i,j}^{S} * A_{i,j-1}^{E}}{L_{i,j-1}^{C}} \tag{7}$$

$$C_{i,j}^{NW} = \frac{A_{i,j}^{W} * A_{i-1,j}^{N}}{L_{i-1,j}^{C}} \tag{8}$$

whereby the diagonals of L and U are named correspondingly to A, and where the subscripts i, j refer to the location of the corresponding grid point. The equation for the L^{C}-entries seem to enforce a sequential computation due to the recursion. An investigation of the data dependencies shows that the parallelization can be done in a pipeline manner when the computational grid is divided into strips and each strip is assigned to one processor. In Figure 1 this technique is illustrated for 3 processors. Grid line segments with the same symbol can be computed simultaneously.

In the following, we denote the side length of the (assumed square) grid by n and the number of processors by p. Dividing the rectangular grid into p vertical strips, each processor p_m is assigned the grid points (i, j) with $imin_m \leq i \leq imax_m$, $1 \leq j \leq n$ and $imax_m - imin_m = n/p + 1$.

Note that the computation of the coefficients L^{C} may be performed with fewer operations by substituting the appropriate values of U^{E} (U^{N}) for the terms A^{E}/L^{C} (A^{N}/L^{C}) in equation (5).

The parallel decomposition algorithm performed by the m-th processor is given in Figure 2, where only the main diagonal coefficients L^{C} are considered. Processors at the end

$$
\boxed{
\begin{array}{l}
\textbf{PROCEDURE Parallel Decomposition} \\
\quad \texttt{FOR } j \texttt{ := 1 TO } n \texttt{ DO} \\
\qquad \texttt{receive from left}(U_{imin_m-1,j}) \\
\qquad \texttt{FOR } i \texttt{ := } imin_m \texttt{ TO } imax_m \texttt{ DO} \\
\qquad\quad L^C_{i,j} := A^C_{i,j} - A^W_{i,j} * U^E_{i-1,j} - A^S_{i,j} * U^N_{i,j-1} \\
\qquad \texttt{send to right}(U_{imax_m,j}) \\
\quad \texttt{END}
\end{array}
}
$$

Figure 2: Standard Parallel ILU-Factorization

$$
\boxed{
\begin{array}{l}
\textbf{PROCEDURE Parallel } k\textbf{-Decomposition} \\
\quad \texttt{FOR } jj \texttt{ := 1 TO } n \texttt{ STEP } k \texttt{ DO} \\
\qquad \texttt{receive from left}(U_{imin_m-1,l}, \quad l = jj, \ldots, jj + k - 1) \\
\qquad \texttt{FOR } j \texttt{ := } jj \texttt{ TO } jj + k - 1 \texttt{ DO} \\
\qquad \texttt{FOR } i \texttt{ := } imin_m \texttt{ TO } imax_m \texttt{ DO} \\
\qquad\quad L^C_{i,j} := A^C_{i,j} - A^W_{i,j} * U^E_{i-1,j} - A^S_{i,j} * U^N_{i,j-1} \\
\qquad \texttt{send to right}(U_{imax_m,l}, \quad l = jj, \ldots, jj + k - 1) \\
\quad \texttt{END}
\end{array}
}
$$

Figure 3: Parallel k-method

of the array ignore the appropriate `send/receive` statement. Note that the algorithm is recursive due to equations (4) and (6).

The parallel algorithm therefore requires $2n$ communications per processor, since grid lines are processed individually. This very fine granularity will lead to poor efficiencies on parallel machines with high message latency.

3. Modifications of the ILU-factorization

The first modification of the factorization algorithm to reduce the number of communications is straightforward. Instead of each processor p_m computing only one line segment of grid points ($imin_m \leq i \leq imax_m, j$) and sending the corresponding U^E-values to its right neighbour p_{m+1}, a block of k line segments ($imin_m \leq imax_m, jmin \leq j \leq jmin + k - 1$) may be computed before a communication takes place (see Figure 4). This modification of the parallel ILU-implementation as shown in Figure 3 will be referred to as the k-method.

This strategy requires a smaller number of communications ($2n/k$) than the original method and thus the granularity is considerably improved. However, the delay until the rightmost processor begins its computation increases. The optimal number k of computed lines before a data exchange is done will vary on different parallel machines depending on

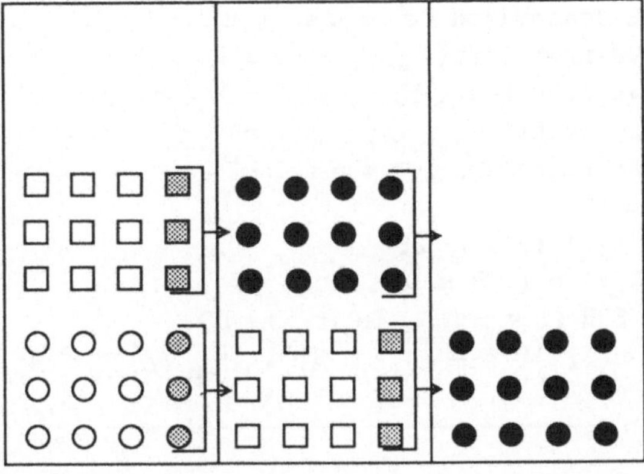

Figure 4: Parallel ILU-factorization with the k-method ($k = 3$)

```
PROCEDURE Modified Parallel Decomposition
    receive from left(U_{imin_m-1,j}, j = 1...n)
    send to right(A^E_{imax_m,j}/A^C_{imax_m,j}, j = 1...n)
    FOR j := 1 TO n DO
        FOR i := imin_m TO imax_m DO
            L^C_{i,j} := A^C_{i,j} - A^W_{i,j} * U^E_{i-1,j} - A^S_{i,j} * U^N_{i,j-1}
END
```

Figure 5: Modified Parallel ILU-factorization

the hardware parameters and the communication software of the processor interconnect. These are essentially the message latency and the channel bandwidth.

Inspection of equation (3) shows that the pipelined computation scheme for L^C is enforced by the recursion which is expressed algorithmically as the dependency on U^E-values from the left neighbour. To overcome this data dependency, the following modification of the factorization may be considered: Each processor p_m uses the A^C-values instead of the L^C-values to compute the entries of U^E in its rightmost column $imax_m$ and communicates these values to its right-hand neighbour p_{m+1}. The factorization of the matrix as described in Figure 5 then makes use of these approximate U^E-values at the western (internal) boundaries $i = imin_m$. After the first communication no further data exchange is needed during the factorization algorithm. We refer to this modification as the *modified* method. With this modification, the total number of messages sent depends no longer on the grid size but only on the number of processors used. This advantage must be paid for by an additional division to compute the approximate values of U^E at the eastern boundary of each processor. The total volume of data transferred remains, however, the same.

All these modifications suffer still from the fact that the factorization is not vectorizable. This is due to the recursion involved in computing the L^C-values. However, if we extend the idea used in the modified algorithm 5 which replaces L^C values by A^C-values in

```
PROCEDURE Vectorized Decomposition
    receive from left(A^C_{imin_m-1,j}, j = 1...n)
    receive from left(A^E_{imin_m-1,j}, j = 1...n)
    send to right(A^C_{imax_m,j}, j = 1...n)
    send to right(A^E_{imax_m,j}, j = 1...n)
    L^{C(0)} := A^C
    FOR s := 0 TO s_max DO
        FOR j := 1 TO n DO
            FOR i := imin_m TO imax_m DO
                L^{C(s+1)}_{i,j} := L^{C(s)}_{i,j} - A^W_{i,j} * A^E_{i-1,j}/L^{C(s)}_{i-1,j} - A^S_{i,j} * A^N_{i,j-1}/L^{C(s)}_{i,j-1}
END
```

Figure 6: Vectorized ILU-Factorization

equation (3) to compute the entries of L^C at *all* grid points, the recursion will vanish and the computation becomes fully vectorizable. Note that this represents a modification of the decomposition and that a corresponding variation in numerical behaviour must be expected. The number of arithmetic operations may not be reduced by the use of the U^E-values instead of the A^E/L^C-term. In Figure 6 we denote the s-th approximation of L^C by $L^{C(s)}$.

Figure 6 thus requires four communications, regardless of the number of processors and may be performed fully in parallel. For the model problem no communication is required, since the coefficients A^C and A^E at the positions $imin_m - 1, j$ are known to processor p_m.

4. Properties of the Vectorized Decomposition

The vectorized algorithm was introduced to enable vectorization and more efficient parallelization on MIMD multiprocessors. However, since a substantial modification has been made, we must re-establish the well-known properties of the original ILU-decomposition. First we analyze the behaviour of the coefficients $L^{C(s)}$ of the vectorized decomposition.

Theorem 1 *The coefficients of the iterated vectorized decomposition converge to those of the original ILU-scheme.*

Proof (Hackbusch): Replace $\leq$ in equation (8.5.18b) by $<$ in the proof to Satz 8.5.10 in [3]. □

In practice, however, it is sufficient to perform one iteration only ($s_{max} = 1$ in Figure 6). In this case the arithmetic cost is equal to that of the original decomposition defined by the equations (1) to (6).

For the use of the vectorized decomposition in a multigrid context we next establish the following important property:

Table 1: Approximate hardware parameters of the machines tested

	α (μs)	β (μs)
Transputer	20	0.71
SUPRENUM	1000	0.1

Theorem 2 *Assume A is spd. The iteration S defined by the vectorized incomplete decomposition on the rectangular grid of mesh size h has the smoothing property*

$$\|AS^\nu\| \leq h^{-\gamma} C \eta(\nu) \tag{9}$$

with

$$\eta(\nu) = max\left\{\frac{\nu^\nu}{(\nu+1)^{\nu+1}}, 2\omega(2\omega-1)^\nu\right\}$$

if the iteration is damped with $0.5 < \omega < 1$.

Proof: See Wittum [7] Theorem 3.1.3. $\square$

As the following theorem shows, the undamped vectorized decomposition has the smoothing property when applied to the model problem.

Theorem 3 *Let A be a symmetric M-matrix arising from the 5-point discretization stencil. Then the vectorized ILU-decomposition has the smoothing property (9) for $\omega = 1$.*

Proof: See Wittum [8] Satz 10. $\square$

5. Experimental Results

In order to investigate the numerical behaviour and the parallel efficiency of the modified ILU-factorizations we implemented them on two parallel machines. The first was a Transputer system (the Meiko Computing Surface [5]) and the second was the SUPRENUM system [6]. In each case up to 16 processors were tested and the program code used was as identical as was possible. The communication overhead of a MIMD machine is characterized by the message latency α and the transfer time per byte β. The values of the parameters for each machine used are given in Table 1.

We use the standard measures for the effectiveness of a parallel method, the Speedup $S(p)$ and the Efficiency $E(p)$, defined as follows:

$$S(p) = \frac{T(1)}{T(p)}, \qquad E(p) = \frac{S(p)}{p}$$

where $T(p)$ represents the computation time for the solution on p processors. In the monoprocessor case we used the time needed for the original ILU-factorization.

Figure 7 shows the efficiencies obtained by the k-method on the SUPRENUM system. The factorization was done on a 200×200 grid. It can be seen that the efficiency improves

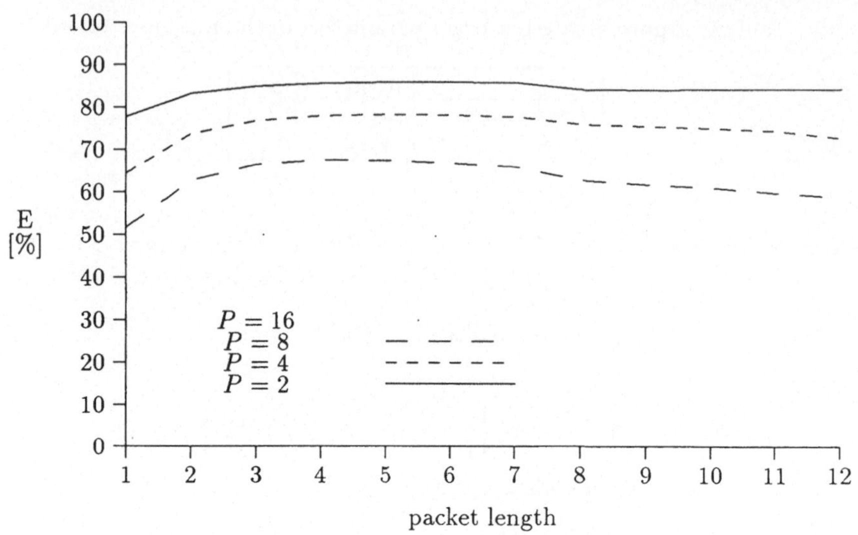

Figure 7: SUPRENUM Efficiencies for the k-method

Table 2: Efficiencies of the modified method

	No. of procs.	2	4	8	16
SUPRENUM	*mod*	0.99	0.99	0.98	0.92
	ilu	0.75	0.58	0.46	0.34
Transputer	*mod*	0.98	0.97	0.96	0.93
	ilu	0.99	0.97	0.94	0.86

for $k \leq 4$, indicating that the message latency is greater than the computation time for the corresponding grid segments. For higher values of k, however, the efficiency decreases as the parallel portion of the algorithm becomes smaller.

Figure 8 shows the results obtained by the implementation of the k-method (Algorithm 3) on the Transputer system. The efficiencies of the standard method ($k = 1$) are seen to be very high, and they deteriorate for larger values of k. This is due the fact that the message latency of the Transputer using the CS_Channel communication primitives is comparatively low and thus the gain from the reduction in number of messages is less than the losses incurred by the increased sequential portion of the algorithm.

Table 2 shows the efficiencies obtained by the modified method on both parallel machines. It can be observed that whilst the efficiencies of the standard method on the Transputer are relatively high, they deteriorate substantially on the SUPRENUM system. The Transputer efficiency is not sensitive to the fine granularity as it has a low message latency. The opposite is true for SUPRENUM (see Table 1).

We implemented the vectorized factorization on the SUPRENUM multiprocessor. The results are presented in Table 3. The grid size was again 200 × 200. In order to improve the accuracy of the time measured, ten factorizations were performed. For comparison

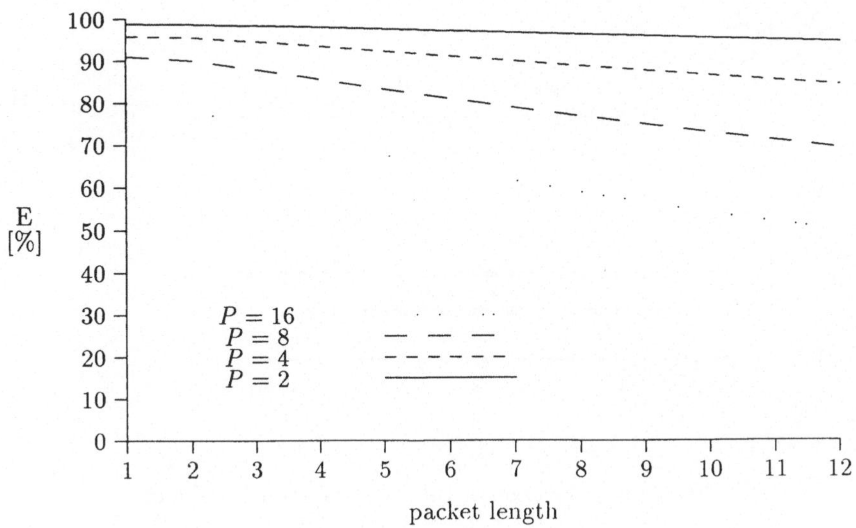

Figure 8: Transputer efficiencies for the k-method

Table 3: Efficiencies of the vectorized factorization on SUPRENUM

	$S(p)$	$E(p)$	$T(p)$	$T(p)$
1	1.00	100.0	3.072	45.98
2	2.00	100.0	1.536	30.64
4	3.99	99.87	0.769	19.72
8	7.96	99.48	0.386	12.38
12	11.68	97.34	0.263	9.392
16	15.21	95.05	0.202	8.556

reasons the timings of the original factorization are given in the last column. It can be clearly seen that there is only a slight reduction in efficiency even on 16 processors. Since there is is no communication involved in this version of the algorithm, the loss in efficiency is solely due to the reduction in vector length when the number of processors is increased. Compared to the original version of the factorization an enormous decrease of computation time can be observed. This is due to the poor scalar performance of the SUPRENUM multiprocessor compared to its vector processing speed.

Figure 9 shows the rate of convergence ρ obtained by a multigrid method using the various ILU versions and different values for the pre- and post-smoothing parameters ν_1 and ν_2. The computation was performed on a finest grid of size 64 $\times$ 64. Bilinear interpolation and full weighting were used for the interpolation and restriction operations, respectively. It can be seen that the vectorized and the modified ILU-factorizations achieve better convergence rates than the original scheme. Note that the original and the vectorized methods are invariant w.r.t. the number of processors, in contrast to the modified scheme. The Figure shows, however, that the numerical behaviour of the latter method is, in practice, independent of p.

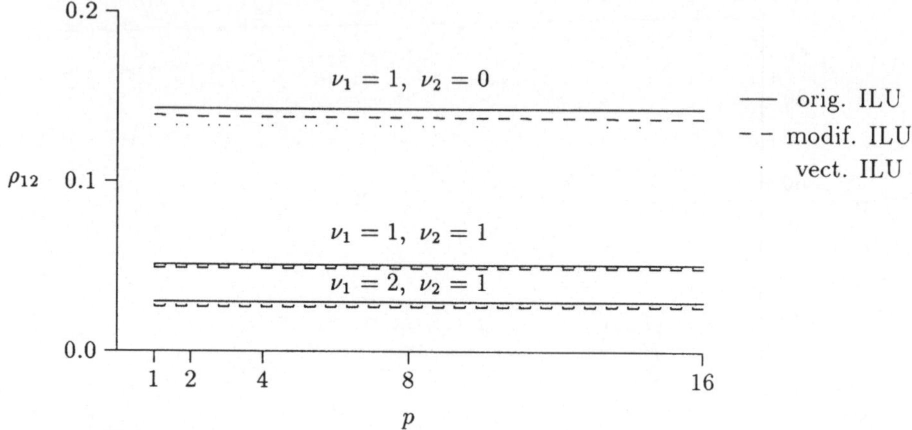

Figure 9: Convergence rates for the model problem

References

[1] P. Bastian, G. Horton : *Parallelization of Robust Multi-Grid Methods: ILU Factorization and Frequency Decomposition Method*. Proceedings of the Fifth GAMM Seminar, Kiel, 1989, Ed. W. Hackbusch and R. Rannacher, Vieweg Verlag, Braunschweig, 1990.

[2] M. J. Flynn : *Some Computer Organizations and their Effectiveness*. IEEE Transactions on Computers, Vol. C-21, No. 9 (Sept. 1972).

[3] W. Hackbusch : *Iterative Lösung großer schwachbesetzter Gleichungssysteme*. Teubner, Stuttgart, 1991.

[4] P. Manneback, J. Qin, G. Libert : *Performance Models of Modified Incomplete Cholesky Conjugate Gradient algorithm on Distributed Memory MIMD Computers*. Proceedings of the European Workshop on Parallel Computing, Barcelona, March, 1992.

[5] Meiko Scientific : *Computing Surface User Manual*, Bristol 1990.

[6] U. Trottenberg : *Contributions of the 2nd International SUPRENUM Colloquium in Parallel Computing*, Parallel Computing 7, 1988.

[7] G. Wittum : *Linear iterations as smoothers in multigrid methods: Theory with applications to incomplete decompositions*. Impact of Computing in Science and Engineering 1 , pp 180 - 215, 1989.

[8] G. Wittum : *Über spektralverschobene Iterationen*. To appear.

Discretization and Iterative Solution
of Convection Diffusion Equations

R.Kornhuber

Konrad–Zuse–Zentrum für Informationstechnik Berlin

Heilbronner Str. 10, D–1000 Berlin 31, Fed. Rep. of Germany

G.Wittum

Institut für Angewandte Mathematik, Universität Heidelberg

Im Neuenheimer Feld 294, D–6900 Heidelberg, Fed. Rep. of Germany

Summary

We propose an extended box method which turns out to be a variant of standard finite element methods in the case of pure diffusion and an extension of backward differencing to irregular grids if only convective transport is present. Together with the adaptive orientation proposed in a recent paper and a streamline ordering of the unknowns, this discretization leads to a highly efficient adaptive method for the approximation of internal layers in the case of large local Peclet numbers.

1. Introduction

We consider boundary value problems for the scalar, linear convection diffusion equation

$$\operatorname{div}(-\varepsilon\nabla u + \beta u) = f \tag{1.1}$$

on a bounded, polygonal domain $\Omega \subset \mathbb{R}^2$ with constant $\varepsilon > 0$ and $\beta = \beta(x) \in \mathbb{R}^2, x \in \Omega$. We assume that β has no loops, i.e. that $\beta(x) \neq \beta(y)$ if $x \neq y$. Additionally, we impose Dirichlet boundary conditions on the inflow boundary $\Gamma_{\text{in}} = \{x \in \partial\Omega | (\beta(x), n) \leq 0\}$ and natural outflow conditions on $\Gamma_{\text{out}} = \partial\Omega \backslash \Gamma_{\text{in}}$:

$$u|_{\Gamma_{\text{in}}} = u_0 , \quad \frac{\partial}{\partial n}u|_{\Gamma_{\text{out}}} = 0. \tag{1.2}$$

Though boundary layers are excluded by (1.2), internal layers still lead to severe problems both with respect to discretization and efficient resolution of the resulting linear system.

In a recent paper KORNHUBER AND ROITZSCH [5] have proposed an adaptive orientation of the underlying triangulation in streamline direction β together with local directed (blue) refinement. Compared to common adaptive strategies they obtained the same or even higher accuracy involving a much lower number of nodes. Note that no a priori information on the flux direction β is used so that the technique carries over to more general situations in fluid dynamics where β is depending on the solution.

The purpose of this paper is to exploit the local structure of the triangulation provided by adaptive orientation and directed refinement, in order to construct a monotone discretization with minimal crosswind. We introduce the so called extended box method which turns out to be a variant of standard Galerkin methods in the self–adjoint case and gives a permanent extension of backward differencing to irregular grids, if only convective transport is present. In the latter case optimal orientation together with streamline ordering of the unknowns leads to a lower diagonal stiffness matrix so that ILU and Gauß–Seidel iterations reduce to exact solvers. Non–vanishing physical or artificial diffusion is viewed as perturbation of this ideal case. It turns out that for large Peclet numbers the resulting linear systems are very efficiently solved by ordinary ILU or Gauß–Seidel iterations while an additional coarse grid transport becomes necessary as soon as the elliptic part of the problem becomes dominant. Note that a related approach to the anisotropic diffusion problem has been carried out successfully by WITTUM [9].

2. The Extended Box Method

Let T be a triangulation of the bounded polygonal domain $\Omega \subset \mathrm{IR}^2$ with $\mathcal{E}$ and $\mathcal{P}$ denoting the set of edges and nodes, respectively. The subset of nodes which are not situated on the Dirichlet boundary is called $\mathcal{P}^0$. We assume that T is regular in the sense that the intersection of two triangles of T is either containing a common edge, a common node or is empty. The finite dimensional space of continuous functions being linear on each $t \in T$ is called $\mathcal{S}(T)$. We provide $\mathcal{S}(T)$ with basis functions $\lambda_p \in \mathcal{S}(T)$, $p \in \mathcal{P}$ defined by $\lambda_p(q) = \delta_{p,q}$ (Kronecker symbol). For a given triangulation T we choose a partition $\mathcal{B} = \{\mathcal{B}_p ,\ p \in \mathcal{P}\}$ with the properties

(B1) $\qquad\qquad \bigcup_{p\in\mathcal{P}} \mathcal{B}_p = \overline{\Omega}$

(B2) $\qquad\qquad \mathcal{B}_p \subset \{t \in T \,|\, p \text{ is vertex of } t\}$

(B3) $\qquad\qquad \mathcal{B}_p \cap \mathcal{B}_q \subset \partial \mathcal{B}_p ,\ p, q \in \mathcal{P}.$

Now integrate the differential equation (1.1) on $\mathcal{B}_q$, $q \in \mathcal{P}^0$, apply Greens formula and use the piecewise linear ansatz

$$U^B = \sum_{p\in\mathcal{P}} U_p \lambda_p \ \in \mathcal{S}(T) \tag{2.1}$$

to obtain the well–known box method

$$\sum_{p\in\mathcal{P}} \left\{ \int_{\partial \mathcal{B}_q} -\varepsilon \frac{\partial}{\partial n_q} \lambda_p + (\beta, n_q)\lambda_p \, d\sigma \right\} U_p = \int_{\mathcal{B}_q} f \, dx \ , \quad q \in \mathcal{P}^0, \tag{2.2}$$

where we used the outward normal n_q to $\partial \mathcal{B}_q$ and the Euclidean scalarproduct $(\cdot, \cdot)$ in IR^2. To avoid instabilities in case of dominating convection, we use a piecewise constant approximation of the first order terms, taking the values $U^B|_{\partial \mathcal{B}_q}$ from upstream. The resulting upwind discretization reads

$$\sum_{p\in\mathcal{P}} \left\{ \int_{\partial \mathcal{B}_q} -\varepsilon \frac{\partial}{\partial n_q} \lambda_p \, d\sigma \right\} U_p +$$

$$+ \left\{ \int_{\partial \mathcal{B}_q} (\beta, n_q)_+ \, d\sigma \right\} U_q + \sum_{p \neq q \in \mathcal{P}} \left\{ \int_{\partial \mathcal{B}_q \cap \partial \mathcal{B}_p} (\beta, n_q)_- \, d\sigma \right\} U_p = \tag{2.3}$$

$$= \int_{\mathcal{B}_q} f \, dx \ , \quad q \in \mathcal{P}^0.$$

with $r_+ = \max(0, r)$ and $r_- = \min(0, r)$, $r \in \mathrm{IR}$, respectively.

It is well–known from the pioneering work of BANK/ROSE [1] and HACKBUSCH [3] that for self–adjoint problems the box method and traditional Ritz–Galerkin scheme are closely related. In fact, if the regularity condition

(B4) $\qquad \partial \mathcal{B}_p \cap e \subset \{ \text{ midpoint of } e \}, \quad p \in \mathcal{P}, e \in \mathcal{E},$

is satisfied, the resulting stiffness matrices coincide and the difference of the approximations U^B and U^G can be estimated by

$$\|U^B - U^G\|_1 \leq \frac{h}{\varepsilon\sqrt{2}}\|f\|_0 \qquad (2.4)$$

with h denoting the minimal size of $e \in \mathcal{E}$ and $\|\cdot\|_1, \|\cdot\|_0$ the usual norms in H^1 and L_2, respectively. Moreover, if f is smooth enough and $\mathcal{B}$ satisfies an additional symmetry condition, related second order estimates are available. We refer to [3] for details.

Obviously the construction of special box–schemes amounts to a special choice of the box–mesh $\mathcal{B}$. In our case this choice is preceded by a suitable extension of the underlying triangulation $\mathcal{T}$. More precisely, from a given $\mathcal{T}$ we construct an extended triangulation $\overline{\mathcal{T}}$ by adding the barycenter s of each triangle $t \in \mathcal{T}$ to the set of nodes and joining it to the vertices as shown in Figure 2.1. Now the corresponding extended box–mesh $\overline{\mathcal{B}}$ is obtained from the subdivision of t in four similar subtriangles, where the inner triangle forms the box $\mathcal{B}_s$ and the others contribute to the boxes related to the remaining vertices.

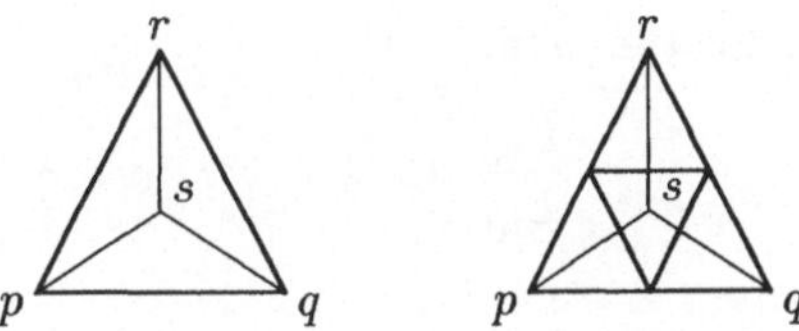

Figure 2.1 Extended triangulation $\overline{\mathcal{T}}$ and box mesh $\overline{\mathcal{B}}$

Note that the additional unknowns U_s are coupled only with U_p, U_q and U_r. Now the extended box method reads as follows.

- Use the upwind box scheme with respect to the extended triangulation $\overline{\mathcal{T}}$ and the related box–mesh $\overline{\mathcal{B}}$.

- Eliminate the additional unknowns U_s by local condensation to obtain the linear system

$$A\underline{U} = b \qquad (2.5)$$

for the remaining unknowns $\underline{U} = (U_p)_{p \in \mathcal{P}^0}$.

We will investigate the properties of this scheme for pure diffusion and pure convection starting with the self–adjoint case $\beta \equiv 0$. Note that the pair $\overline{\mathcal{T}}, \overline{\mathcal{B}}$ does not satisfy the regularity condition (B4). However, we can state the following Theorem which is closely related to the results summarized above.

Theorem 2.1 *Let $\beta \equiv 0$. Then A coincides with the stiffness matrix resulting from the standard Ritz – Galerkin Method with respect to the original triangulation T. Moreover, the difference of U and the Ritz – Galerkin approximation U^G can be estimated according to (2.4).*

Proof: Let us apply the standard box method to the extended meshes $\overline{T}$ and $\overline{B}$. In view of (2.2), the contributions of some triangle $t \in T$ to the stiffness matrix $\overline{A}$ are given by

$$\overline{L}_{\nu,\mu} = -\int_{\partial \overline{B}_\nu \cap t} \varepsilon \frac{\partial}{\partial n_\nu} \overline{\lambda}_\mu \, d\sigma, \quad \nu, \mu = p, q, r, s, \tag{2.6}$$

with $\overline{\lambda}_\mu, \mu \in \overline{\mathcal{P}}$ denoting the standard basis functions in $\mathcal{S}(\overline{T})$. Note that $\overline{L}_{s,\mu}, \mu = p, q, r$, coincides with the entries of $\overline{A}$ as $\overline{B}_s$ is contained in t. Hence the local contributions $L_{\nu,\mu}$ to A can be obtained by local condensation:

$$L_{\nu,\mu} = \overline{L}_{\nu,\mu} - \frac{\overline{L}_{s,\mu}}{\overline{L}_{s,s}} \overline{L}_{\nu,s} , \quad \nu, \mu = p, q, r. \tag{2.7}$$

To show the desired equality of the stiffness matrices, we will prove

$$L_{\nu,\mu} = \int_t \varepsilon(\nabla\lambda_\nu, \nabla\lambda_\mu) \, dx =: L^G_{\nu,\mu} , \quad \nu, \mu = p, q, r. \tag{2.8}$$

For this reason assume for the moment that

$$\frac{\overline{L}_{s,\mu}}{\overline{L}_{s,s}} = -\tfrac{1}{3} , \quad \mu = p, q, r. \tag{2.9}$$

Then in view of

$$\overline{\lambda}_\nu + \tfrac{1}{3}\overline{\lambda}_s = \lambda_\nu \quad \text{on } t, \quad \nu = p, q, r, \tag{2.10}$$

the desired identity (2.8) takes the form

$$-\int_{\overline{B}_\nu \cap t} \varepsilon \frac{\partial}{\partial n_\nu} \lambda_\mu \, d\sigma = L^G_{\nu,\mu} , \quad \nu, \mu = p, q, r. \tag{2.11}$$

As $\partial \overline{B}_\nu$ intersects the edges of t at their midpoints, (2.11) is easily shown along the lines of [1] or [3]. We still have to prove (2.10) which can be rewritten as

$$\int_{\partial \overline{B}_s} \frac{\partial}{\partial n_s}(\overline{\lambda}_\mu + \tfrac{1}{3}\overline{\lambda}_s) \, d\sigma = 0. \tag{2.12}$$

Now in view of (2.11) the assertion easily follows from Greens formula. To estimate the difference of the approximations U and U^G, observe that on each triangle $t \in T$ the right hand side is approximated by

$$F_\nu = \int_{\overline{B}_\nu} f \, dx + \alpha_\nu \int_{\overline{B}_s} f \, dx , \quad \nu = p, q, r, \tag{2.13}$$

with coefficients α_ν given by

$$\alpha_\nu := \frac{\overline{L}_{\nu,s}}{\overline{L}_{s,s}} = \frac{|e_\nu|}{|e_p| + |e_q| + |e_r|} , \quad \nu = p, q, r. \tag{2.14}$$

Now the proof is completed following almost literally the arguments in [3].

∎

If $\mathcal{T}$ satisfies the strong regularity condition $|e| = $ const., $e \in \mathcal{E}$, then (2.13) obviously holds with $\alpha_\nu = \frac{1}{3}, \nu = p, q, r$, and the extended box method reduces to a variant of the box method proposed by HACKBUSCH[3]. In this case we have an analogue of (2.4) of second order. Note that for $\beta \neq 0$ the right hand side f is still approximated as in (2.13) but with α_ν replaced by a modified partition of unity α_ν^β, $\nu = p, q, r$.

Let us now consider the case of pure convection appearing as reduced problem of (1.1) in the sense of asymptotic analysis.

Theorem 2.2 *Let $\varepsilon = 0$. Assume that $\beta(x) = $ const., $x \in \Omega$ and that $\mathcal{T}$ is oriented in the sense that in each triangle $t \in \mathcal{T}$ one edge is parallel to β. Then the extended box method reduces to the first order upwind difference scheme.*

Proof: Let $p \in \mathcal{P}^0$. Then it is easily seen from the assumptions on $\mathcal{T}$ and the boundary conditions (1.2) that there is one and only one edge $e_0 = (p, p_0) \in \mathcal{E}$ pointing from p in upstream direction. Hence upwind differencing is feasible. Refering to the denotation introduced in Figure 2.1, we first observe that in case of pure convection the unknowns U_p, U_q, and U_r are coupled only via the intermediate unknowns U_s. Exploiting the orientation of $\mathcal{T}$ in (2.3), it is easily checked that U_s and U_q or U_r are coupled only if $q = p_0$ or $r = p_0$.

∎

The above Theorem makes sure that for oriented triangulations there is no unphysical coupling of different streamlines. This is not the case in usual variants of the Petrov–Galerkin method (see [4]), where certain averages of upwind differences are obtained. Of course, there is no artificial viscosity in this ideal case, while on unstructured grids we expect the usual diffusive behaviour of an upwind method.

Of course, the condition $\beta(x) \equiv$ const. can be weakened.

Having treated the ideal cases $\varepsilon = 0$ and $\beta = 0$ let us finally state that the properties of the extended box method applied to general convection diffusion equations are not obtained as the sum but as a nonlinear interpolation of these results. This is due to the local elimination of the U_s which is not additive.

3. Iterative Solution

In this chapter we consider the iterative solution of the linear system (2.5) resulting from the extended box method. We further assume that the triangulation $\mathcal{T}$ is oriented in the sense of Theorem 2.2 and that $\varepsilon \ll |\beta|$. This allows to treat the general convection diffusion equation as perturbation of the singular case $\varepsilon = 0$.

Theorem 3.1 *Let $\varepsilon = 0$. Assume that $\beta(x) \equiv$ const., $x \in \Omega$, and that $\mathcal{T}$ is oriented in the sense that in each triangle $t \in \mathcal{T}$ one edge is parallel to β. Then the nodes $\mathcal{P}$ can be ordered in streamlines following the flux direction β. With respect to this ordering the extended box method provides a lower triangular stiffness matrix A.*

Proof: The Proof is obvious from Theorem 2.2 .

∎

Again, the assumption $\beta \equiv$ const. can be weakened.

Of course iterative schemes like the ILU or Gauß–Seidel iteration reduce to exact solvers, if the assumptions of Theorem 3.1 are fulfilled. But this ideal situation may be disturbed in several ways. First adaptive orientation of $\mathcal{T}$, as performed in the sequel, is supposed to be not exact. Hence a certain amount of artificial diffusion may be introduced by the discretization. On the other hand, any nonvanishing physical diffusion becomes dominant as soon as the local stepsize is chosen small enough. Both phenomena will show up in our numerical experiments. We conjecture that these problems arising from too much diffusion can be remedied by additional coarse grid correction, using the iterative scheme as smoother. This will be subject to subsequent work.

We are left with the problem to provide an efficient streamline ordering of the unknowns. This will be done in the well–known adaptive framework described for example in [2] or [5]. Starting with some initial triangulation $\mathcal{T}_0$, we assume that a sequence of triangulations $\mathcal{T}_1, \mathcal{T}_2, \ldots, \mathcal{T}_j$ is obtained from local regular (red) or directed (blue) refinement and irregular (green) closure. The enumeration of the corresponding nodes will be performed inductively with respect to the refinement levels j.

First, let us briefly discuss the additional requirements on the underlying data structures. As we do not insist on uniform refinement, the polygonals of edges following the streamlines may be disconnected and grow together on subsequent levels. Hence, the nodes belonging to different streamlines are stored in different linked lists. By a pointer on the initial node of the corresponding list, each node is located in its line. The further implementation heavily relies on the data structures provided by the underlying finite element code KASKADE. In particular we have a pointer from each refined triangle $T \in \mathcal{T}_j$, called father, to the resulting triangles $t_1, t_2, t_3, t_4 \in \mathcal{T}_{j+1}$, called sons, and vice versa. Note that t_3, t_4 are void in case of blue refinement. We refer to LEINEN[6] or ROITZSCH[7, 8] for details.

Now the initial numeration of $\mathcal{P}_0$ corresponding to the intentionally coarse grid $\mathcal{T}_0$ may be provided by the user or derived automatically from the approximation U_0 on level 0. In any case the initial grid has to be fine enough to guarantee that the detected streamline direction does not change dramatically in course of the refinement process (see [5]).

In the induction step we assume that the initial triangles $T \in \mathcal{T}_0$ are ordered according to the numeration of the nodes $\mathcal{P}_0$. Based on the existing numeration of $\mathcal{P}_j$ the new nodes $p \in \mathcal{P}_{j+1} \backslash \mathcal{P}_j$ arising from the actual refinement step are sorted by the recursive procedure NUMERATE($\cdot$) which is applied subsequently to the initial triangles $T_0, T_1, \ldots, T_{N_0} \in \mathcal{T}_0$.

procedure NUMERATE(t)
    while (t has sons t_1, t_2, t_3, t_4)
       if ($t \in \mathcal{T}_{j+1} \backslash \mathcal{T}_j$)
           file the new points into an existing line or make up a new one
      else
           perform NUMERATE(t_i), $i = 1, \cdots, 4$, following the order
           induced by the existing numeration of $\mathcal{P}_j$ (see Figure 3.1) .

The above algorithm may be viewed as a recursive search for new nodes $p \in \mathcal{P}_{j+1} \backslash \mathcal{P}_j$, taking care that they are picked up according to the existing order of $\mathcal{P}_j$. Note that the numeration follows the streamlines together with the corresponding polygonals of edges.

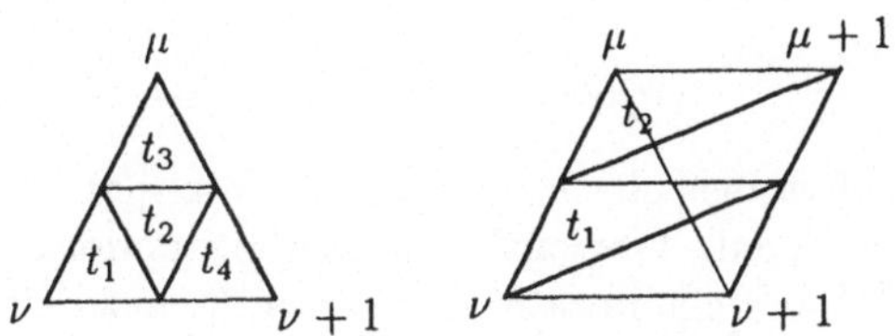

Figure 3.1 Local numeration of sons

Hence local orientation only leads to a local streamline ordering of the nodes.

4. Numerical Results

In our numerical experiments we concentrate on the behaviour of the extended box method and of the ILU iteration applied to the resulting linear systems. For details on the underlying adaptive concept we refer to [5].

In our first example we chose $\beta \equiv (1, 0.5)^T$, $f \equiv 0$ on $\Omega = [0 \times 1, 0 \times 1]$ and the boundary conditions (1.2) with discontinuous u_0 defined by $u_0(x_1, x_2) = 1$ if $x_2 \leq 0.3$ and $u_0 = 0$ elsewhere on Γ_{in}. The initial triangulation $\mathcal{T}_0$ is depicted in Figure 4.1.

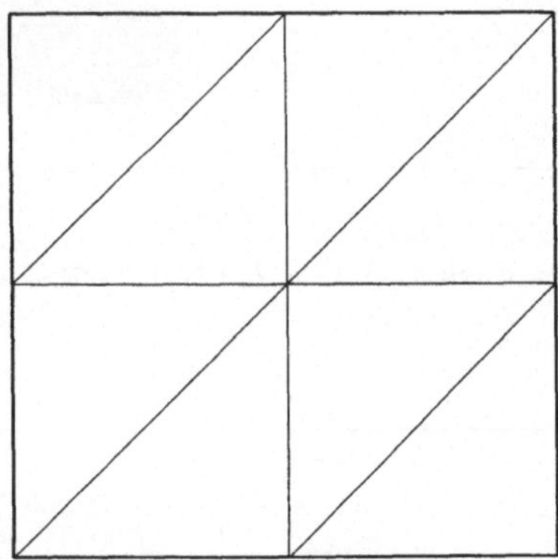

Figure 4.1 Initial triangulation $\mathcal{T}_0$

We first illustrate the properties of the extended box method in the case of pure convection $\varepsilon = 0$. The next Figure 4.2 shows the approximation $\tilde{U}_9$ together with the corresponding triangulation $\tilde{\mathcal{T}}_9$, resulting from 9 isotropic refinement steps without orientation. As expected, we observe the monotone but diffusive behaviour of usual upwind schemes. Using adaptive orientation and directed (blue) refinement, we end up with the anisotropic triangulation $\mathcal{T}_9$ depicted in Figure 4.3. In this case, the extended box method leads to the approximation U_9, introducing almost no unphysical crosswind but preserving the good stability properties. Note that $\mathcal{T}_9$ is involving a much smaller number of nodes than $\tilde{\mathcal{T}}_9$. The next Figure 4.4 shows the convergence properties of the ILU iteration applied to the linear systems (2.5), arising from the extended box method for the choice $\varepsilon = 10^{-4}, 10^{-5}, 0$. We depict the number of iterations depending on the refinement levels which are taken as a measure for the local stepsize h. We always started with $U^{(0)} = 0$ and required the unrealistic accuracy

$$|U^{(\nu+1)} - U^{(\nu)}|_{\ell_2} \leq 10.^{-12}$$

to amplify the effects of various physical diffusion. Note that only the adaptively oriented triangulations are considered. Obviously the method performs excellent for large local Peclet numbers $\rho = \frac{h|\beta|}{\varepsilon}$ and deteriorates as soon as the local mesh size h is small enough to resolve the internal layer. Note that the small increase of the line corresponding to $\varepsilon = 0$ is due to non–optimal orientation. It should be mentioned that the Gauß – Seidel method shows a very similar behaviour but appears to be less robust with respect to perturbing diffusion.

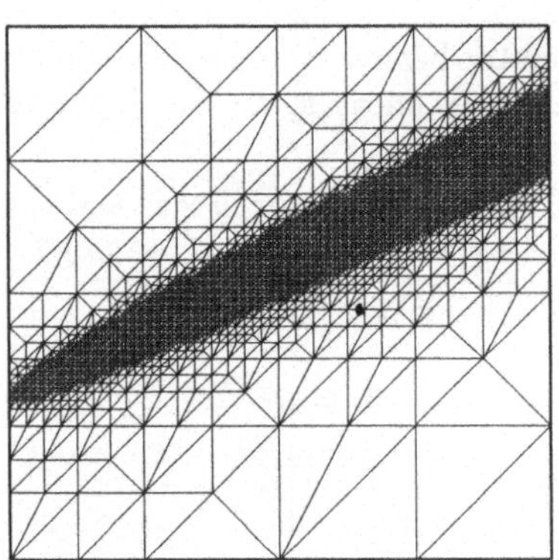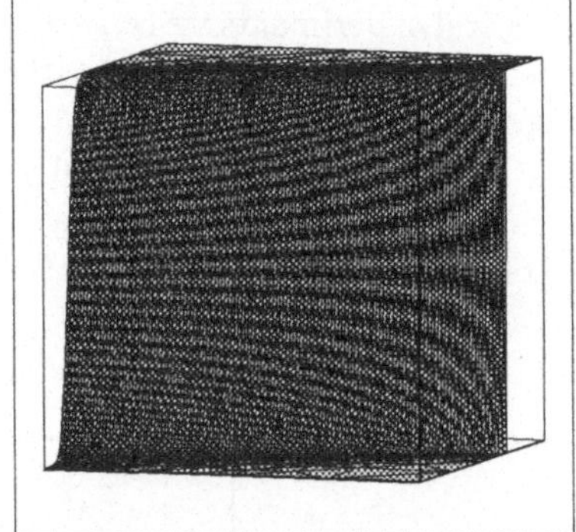

Figure 4.2 Standard triangulation $\tilde{\mathcal{T}}_9$ (2716 nodes) and approximation $\tilde{U}_9$

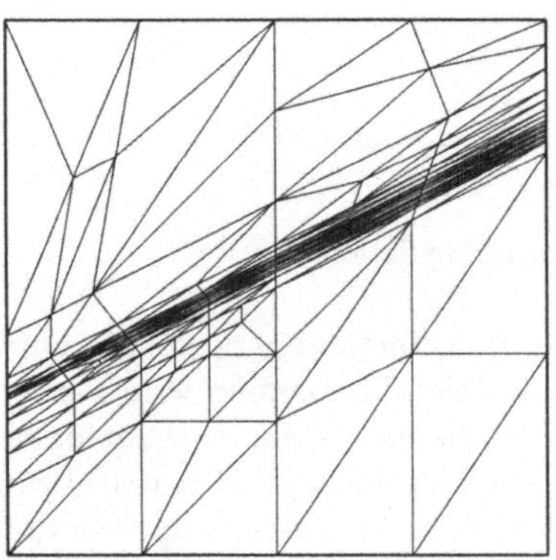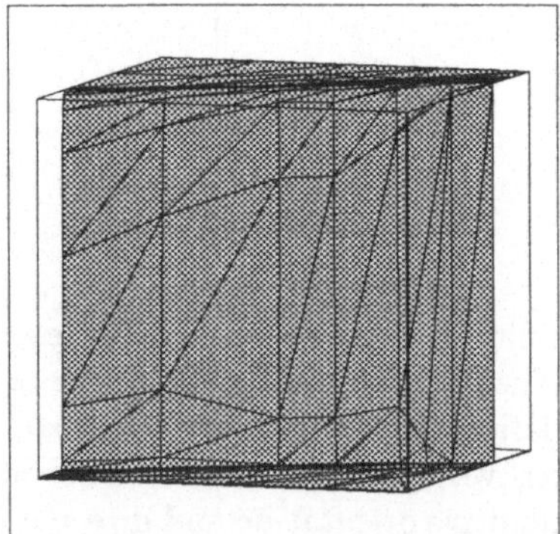

Figure 4.3 Oriented triangulation $\mathcal{T}_9$ (121 nodes) and approximation U_9

In the second example we produce a curved layer by the choice $\beta(x_1, x_2) = (x_2, -x_1)^T$ and a corresponding modification of u_0. The other data including the initial triangulation remain unchanged. Of course we cannot expect that the non–linear layer is approximated by linear edges as good as the linear layer considered above. Indeed, the effect of less

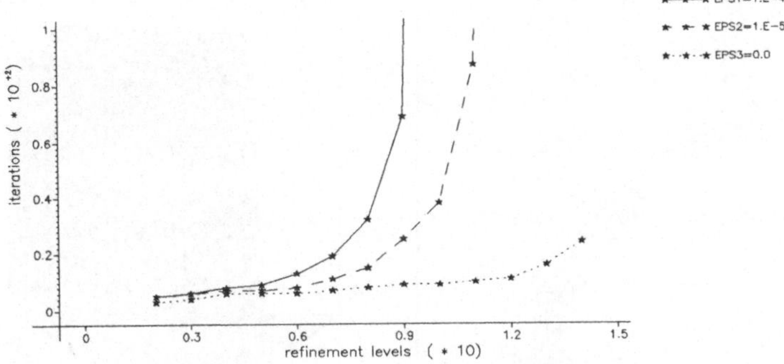

Figure 4.4 ILU iteration for a linear layer

accurate orientation is visible both in the diffusivity of the approximate solution (see Figure 4.5) and the behaviour of the iterative solver (see Figure 4.6). But compared to the solution produced by a standard triangulation and illustrated in Figure 4.4, this approach still provides a considerable advantage both with respect to accuracy and efficiency.

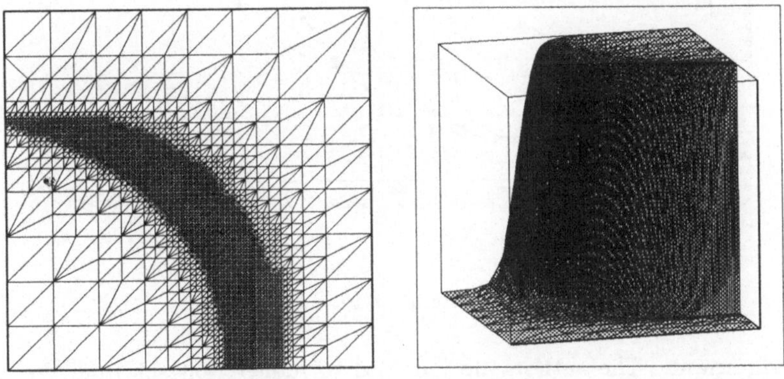

Figure 4.5 Standard triangulation $\tilde{T}_9$ (3106 nodes) and approximation $\tilde{U}_9$

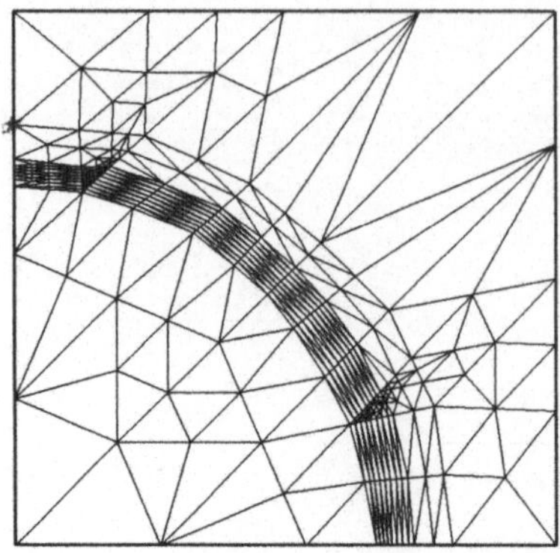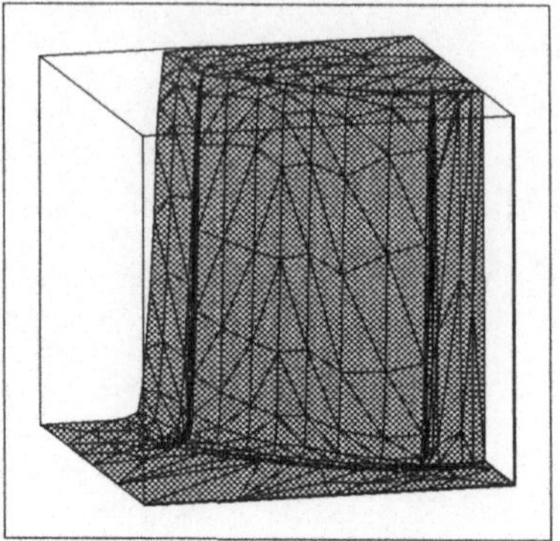

Figure 4.6 Oriented triangulation T_9 (243 nodes) and approximation U_9

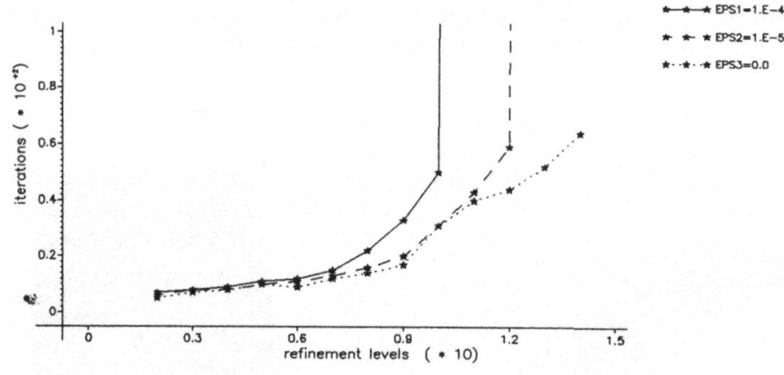

Figure 4.7 ILU iteration for a curved layer

Acknowledgement. The authors are indebted to Rainer Roitzsch from the Konrad–Zuse–Zentrum Berlin for his support at all stages of the preparation of this manuscript.

References

[1] R.E. Bank, D.E. Rose: *Some Error Estimates for the Method.* SIAM J. Numer. Anal. **24** No. 4, p. 777–787 (1987).

[2] P. Deuflhard, P. Leinen and H. Yserentant: *Concepts of an Adaptive Hierarchical Finite Element Code.* IMPACT **1**, p. 3–35 (1989).

[3] W. Hackbusch: *On First and Second Order Box Schemes.* Computing **41**, p. 277–296 (1989).

[4] C. Johnson: *Numerical Solutions of Partial Differential Equations by the Finite Element Method.* Cambridge University Press, Cambridge (1987).

[5] R. Kornhuber, R. Roitzsch: *On Adaptive Grid Refinement in the Presence of Internal or Boundary Layers.* IMPACT **2**, p. 40–72 (1990).

[6] P. Leinen: *Ein schneller, adaptiver Löser für elliptische Randwertprobleme auf Seriell- und Parallelrechnern.* Thesis, University of Dortmund (1990).

[7] R. Roitzsch: *KASKADE User's Manual.* Technical Report TR 89–4, Konrad–Zuse–Zentrum Berlin (ZIB) (1989).

[8] R. Roitzsch: *KASKADE Programmer's Manual.* Technical Report TR 89–5, Konrad–Zuse–Zentrum Berlin (ZIB) (1989).

[9] G. Wittum: *On the Robustness of ILU–Smoothing.* SIAM J. Sci. Stat. Comput. **10**, p. 699–717 (1989).

Empirically modified block incomplete factorizations

MAGOLU monga-Made[1]

Université Libre de Bruxelles
Service de Métrologie Nucléaire (CP 165)
50 av. F.D. Roosevelt
B-1050 Brussels, Belgium
email : mmmago@ulb.ac.be

Summary

Dynamic variants of modified block incomplete factorization with modulated additive perturbations have been introduced recently and found to be superior to the standard block methods, in particular when applied to difficult problems for which the usual modified version gives rise to strongly isolated largest eigenvalues. We show here that in the case where the *PDE* coefficients are strongly anisotropic, care should be taken to provide some enough selection of the perturbed nodes in order to avoid a severe loss of efficiency.

1. Introduction

Consider solving large and sparse symmetric positive (semi)definite *compatible* linear systems

$$Au = b \qquad (b \in \mathcal{R}(A)) \tag{1}$$

by the preconditioned conjugate gradient (*PCG*) method. Let B denote the preconditioner or preconditioning matrix. It is well known that the number of iterations to converge to a specified relative error grows at most like $\mathcal{O}(\sqrt{\kappa})$ where κ stands for the spectral condition number of the pencil of matrices $A - \lambda B$ defined as the ratio of its largest to its smallest *positive* eigenvalue (see e.g. [2] for the regular case and [16] for the singular case). But the rate of convergence also depends on the distribution of the eigenvalues (see e.g. [3, 16, 18]).

In the case where the matrix B represents some *modified* block incomplete factorization of the system matrix A, simple analytical upper bounds on the spectral condition number have been obtained in [9] and in ([13] : Chapter 3 and Appendix B). The results

---

[1] Research supported by the "Programme d'impulsion en Technologie de l'Information" financed by Belgiam State, under contrat Nr. IT/IF/14.

allow to determine *in advance* the order of magnitude of κ for a large class of second order self-adjoint (elliptic) *PDEs*. In particular, it is shown that in *favourable* situations, κ is reduced from $\mathcal{O}(h^{-2})$ to $\mathcal{O}(h^{-1})$ for discrete *PDEs* with (average) mesh parameter h. Earlier results on the topic may be found in [5, 8, 14, 17].

To ensure the same behaviour in unfavourable circumstances, *dynamic* variants of modified incomplete block-matrix factorization with additive perturbations were introduced and succesfully tested in [10, 12]. Our purpose in this work is to illustrate without going into full details that, in the presence of (strong) anisotropy, care should be taken to provide some enough selection of the perturbations introduced by a dynamic factorization method, otherwise they could give rise to degenerate smallest eigenvalues, which results in disappointing performances for perturbed methods compared with unperturbed ones, even when the (effective or reduced) spectral condition number behaves good. By the effective spectral condition number, we understand the spectral condition number computed from the spectrum of the preconditioned system deprived of some of its isolated members (see e.g. [2] and references collected therein).

The outline of the paper is as follows. General terminology and notation are gathered in Section 2. In Section 3, we first recall a particular dynamic block factorization that was introduced and termed *Strategy 2* in [10], and thoroughly examined in [13] under the name of *dynamically modified block incomplete-LU factorization with parameter* α or in abbreviated form, *DMBILU(α)*. We next explain why in the case of (strong) anisotropy, the convergence of the *PCG* method can slow down, which motivates the introduction of what we call *empirically modified block incomplete-LU factorization with parameter* α or briefly, *EMBILU(α)*. By way of illustration, the results of numerical experiments are presented in Section 4.

2. Terminology and notation

The symbols A^t, A^+, $\mathcal{N}(A)$, $\mathcal{R}(A)$, $\lambda_{min}(A)$, $\lambda_{max}(A)$ stand for respectively the transpose, the Moore-Penrose inverse [6], the null space, the range, the smallest and the largest *nonzero* eigenvalues of the matrix A. By $\mathcal{P}_{S,T}$ we denote the projector with range S and null space T (S and T are complementary subspaces). We use e to denote the vector whose all components are equal to 1. A matrix whose nonzero entries are equal to 1 is referred to as a $(0,1)$ *matrix*.

Hadamard multiplication The Hadamard product $A*B$ of the matrices $A = (a_{ij})$ and $B = (b_{ij})$ of the same dimensions is the element by element multiplication $(A * B)_{ij} = a_{ij}b_{ij}$. The unit matrix with respect to the Hadamard multiplication, denoted ε, is the matrix whose entries are all equal to unity.

Partitionings Any partitioning of an n-vector $x = (x_I)$ into block components x_I of dimensions $n_I, I = 1, 2 \ldots, M$ (with $\sum_{I=1}^{M} n_I = n$) is uniquely determined by a partitioning $\pi = (\pi_I)_{1 \leq I \leq M}$ of the set of the first n integers. We assume throughout this work that all n-vectors are partitioned in blocks according to a given such partitioning. The same partitioning π induces also a partitioning of any $n \times n$ matrix A into block components

A_{IJ} of dimensions $n_I \times n_J$ and we shall similarly assume that all $n \times n$ matrices are partitioned in this way. Lower case indices refer to scalar entries and capital indices to block entries. Thus scalar (resp. block) entries of an $n \times n$ π-partitioned matrix A are denoted a_{ij} (resp. A_{IJ}). When needed, scalar entries of block entries of A are denoted $(A_{IJ})_{ij}$, a notation which implies that $i \in \pi_I$ and $j \in \pi_J$. Similar notations are used for vector components, except that we always represent vectors by small letters.

A matrix which is block diagonal (resp. tridiagonal, triangular) relative to the π-partitioning is referred to as π-diagonal (resp. π-tridiagonal, π-triangular). We use the notation $diag_\pi(A)$ to denote the π-diagonal matrix with block diagonal equal to that of A, and $offdiag_\pi(A)$ to denote $A - diag_\pi(A)$. In the case of the usual point partitioning, we use $\pi = p$.

Stieltjes matrix A real square matrix A is called a *Stieltjes matrix* if it is symmetric positive (semi)definite and if all its offdiagonal entries are not positive.

Standard point LU-factorization By the *standard point LU-factorization* of a (Stieltjes) matrix P, we understand the factorization $P = U^t P^+ U$ such that U is p-upper triangular and $P = diag_p(U)$.

3. Dynamically modified block incomplete factorizations

For simplicity, we shall confine ourself to π-tridiagonal Stieltjes matrices. Let A denote such a given matrix such that

(a) each π-diagonal entry A_{II} of A is irreducible,

(b) for $I = 1, 2, \cdots, M - 1 \qquad A_{I,I+1} \neq 0$.

Assume further that β and γ are specified $(0,1)$ matrices while α denotes some given parameter such that $0 < \alpha < 1$. Let x be a positive vector such that $Ax \geq 0$ and let P denote the π-diagonal matrix whose entries are computed according to the following algorithm

For $i \in \pi_1$ compute

$$(\Delta_{11})_{ii} = \max\left\{0, \frac{(A_{12}x_2)_i}{1-\alpha} - (A_{11}x_1)_i\right\} \frac{1}{x_i}$$

$P_{11} = A_{11} + \Delta_{11}$

For $I = 2, 3, \cdots, M$ compute

$$K_{I-1,I-1} = \gamma_{I-1,I-1} * P_{I-1,I-1}^{-1}$$

$$P_{II}^{(0)} = A_{II} - \beta_{II} * (A_{I,I-1}K_{I-1,I-1}A_{I-1,I}) - \Omega_{II}$$

where Ω_{II} is the p-diagonal matrix defined by

$$\Omega_{II}x_I = \left(A_{I,I-1}P_{I-1,I-1}^{-1}A_{I-1,I} - \beta_{II} * (A_{I,I-1}K_{I-1,I-1}A_{I-1,I})\right) x_I$$

For $i \in \pi_I$, $I < M$ compute

$$(\Delta_{II})_{ii} = \max\left\{0, \frac{(A_{I,I+1}x_{I+1})_i}{1-\alpha} - (P_{II}^{(0)}x_I)_i\right\} \frac{1}{x_i}$$

$$P_{II} = P_{II}^{(0)} + \Delta_{II} \qquad (\Delta_{MM} = 0) .$$

Let F denote the strictly π-upper triangular matrix whose nonzero block entries are given by $F_{I,I+1} = A_{I,I+1}$ for $I = 1, 2, \cdots, M - 1$. The matrix $B = (P - F^t)P^{-1}(P - F)$ (which is well defined) is referred to as the *dynamically modified block incomplete LU-factorization with parameter α (DMBILU(α))* of A associated with x, β and γ [13]. This technique of factorization was termed *Strategy 2* in [10]. The matrices A and B satisfy the relation $Bx = Ax + \Delta x$, the p-diagonal matrix Δ is called the perturbation matrix.

With $\alpha = h/s$, where s denotes a $\mathcal{O}(1)$ parameter, one has that $\lambda_{max}(B^{-1}A) \leq sh^{-1}$ [10, 13], in other words the technique allows to keep control of $\lambda_{max}(B^{-1}A)$ and implicitly avoid strong isolation of the largest eigenvalues. Moreover, the number of *PCG* iterations are weakly influenced by the choice of the parameter s, provided that it is chosen around $1/4$ (more specifically $\frac{1}{8} \leq s \leq 1$). On the other hand, if the p-diagonal entries of A are all positive and if the coefficients of A are normalized so that the p-diagonal entries are $\mathcal{O}(1)$, it can be shown by using techniques developed in [4, 11, 16] that $\lambda_{min}(B^{-1}A) \geq \mathcal{O}(1)$ on the assumption that the number of nodes where $\mathcal{O}(h)$ perturbations are added is (bounded by) $\mathcal{O}(h^{-1})$. Consequently, $\kappa(B^{-1}A) \leq \mathcal{O}(h^{-1})$.

As far as the *PCG* method is concerned, such a behaviour of the spectral condition number, which reveals nothing about the distribution of the smallest eigenvalues, does not guarantee that perturbed methods will always yield better results than unperturbed ones. Indeed, It turns out from theoretical investigations and/or illuminating experimental results by van der Sluis and van der Vorst [18, 19], by Greenbaum [7] and by Notay [16] that the higher the number of (strongly) isolated smallest eigenvalues, the longer the delay for the process to enter into its so-called *superlinear convergence phase*, i.e. the phase during which the method behaves as if these isolated eigenvalues were absent.

Now, it follows from general results on perturbed methods by Axelsson [1] (see [16] for the singular case) that the eigenvalues of $B^{-1}A$ are related to those of the original system matrix A. Further

- The smaller the size of the perturbations and/or the number of perturbed nodes, the stronger the clustering of the smallest eigenvalues around 1.

- The larger the size of the perturbations and/or the number of perturbed nodes, not only the stronger the isolation of the smallest eigenvalues from 1 but also the stronger the similarity between the distribution of the smallest eigenvalues of B^+A with that of the original system matrix A.

Now, in the presence of anisotropy, the number of strongly isolated smallest eigenvalues of the original PDE matrix A, thus also for $B^{-1}A$, increases (see e.g. [1]), which could cause the delay (measured in number of iterations) needed by the PCG process to eliminate the components of the initial error vector along the corresponding eigenvectors to get longer, with the possibility of annihilating the benefit from the superlinear convergence effects.

To prevent such a loss of efficiency, we advocate to preclude dynamic preconditionings from perturbing in regions where the PDE coefficients are (strongly) anisotropic. Within our framework, unsafe nodes may be characterized by

$$\max\left\{(A_{I,I+1}e_{I+1})_i, (H_{II}e_I)_i\right\} \geq \psi \min\left\{(A_{I,I+1}e_{I+1})_i, (H_{II}e_I)_i\right\} \qquad i\in\pi_I, \ I<M \quad (2)$$

where $H_{II} = -offdiag_p(A_{II})$ and $\psi > 1$ denotes some tolerance setting. Numerical experiments performed over a wide range of $2D$ problems indicate values of ψ larger than or equal to (resp. much smaller than) 10 as good choices for difficult (resp. easy) problems. By *easy* anisotropic problems we understand $PDEs$ for which the coefficients are anisotropic (almost) everywhere in the domain. In such a situation, the unperturbed modified block methods are already so performant that there is no serious reason to add perturbations; this often makes things worse.

The empirical version of $DMBILU(\alpha)$ where perturbations are set to zero at nodes i for which the inequality (2) holds with $\psi = 10$, will be referred to as the *empirically modified block incomplete LU-factorization with parameter* α or in abbreviated form $EMBILU(\alpha)$.

4. Numerical results

To check the validity of our argument, we consider the following two-dimensional PDE

$$-\frac{\partial}{\partial x}\left(p\frac{\partial}{\partial x}u(x,y)\right) - \frac{\partial}{\partial y}\left(q\frac{\partial}{\partial y}u(x,y)\right) = f(x,y) \quad \text{in } \Omega = (0,1)\times(0,1)$$

$$\frac{\partial}{\partial n}u(x,y) = 0 \qquad \text{on } \partial\Omega$$

with the compatibility condition

$$\int_\Omega f(x,y)\,dx\,dy = 0 \ .$$

We assume the domain to be subdivided as depicted in Fig. 1 where the coefficients p and q are also specified. The difficulty lies on the (opposite) presentation of anisotropy in Ω_2 and Ω_3. The mesh size is $h = 1/(M - 1)$. For convenience, $M - 1$ is assumed to be a multiple of 8. We use the five-point finite difference approximation (point scheme box integration); the mesh points are numbered **lexicographically**. The resulting matrix A is a block-tridiagonal irreducibly diagonally dominant (singular) Stieltjes matrix, which we partition according to the line partitioning $(\pi_I)_{1 \leq I \leq M}$, keeping together the unknowns in the x-direction. The right-hand side of the system to solve is chosen such that the function $u_0(x, y) = x(1 - x)y(1 - y)e^{xy}$ generates a solution on the grid. To prevent instability problems, the residual vector as well as the vector solution to the preconditioning system should be projected onto $\mathcal{R}(A)$, <u>at each *PCG* iteration</u> [16].

For comparison purposes, the preconditionings include :

1. **MBILU** : The standard modified block method ($\Delta = 0$).

2. **DMBILU**(α) : with $\alpha^{-1} = sM$, s denoting a $\mathcal{O}(1)$ parameter.

3. **EMBILU**(α) : with $\alpha^{-1} = sM$, s denoting a $\mathcal{O}(1)$ parameter.

All the involved incomplete factorizations are associated with $x = e$, $\beta = tridiag_p(\varepsilon)$ and $K_{II} = tridiag_p(P_{II}^{-1})$ for $I = 1, 2, \ldots, M - 1$. Since A is an irreducible singular Stieltjes matrix that satisfy the requirements *(a)* and *(b)* of Section 4, we know from [12, 13] that its *MBILU* factorization $B = (P - F^t)P^+(P - F)$ is a singular matrix such that only the last block diagonal entry P_{MM} of P is singular; in which case, to solve the preconditioning system, it suffices to replace, during the standard pointwise factorization of P_{MM}, the last p-diagonal entry (which is zero) by 1, which amounts to resorting to a generalized inverse of the preconditioning matrix. Given that $\mathcal{N}(A) = span\{e\}$ (see e.g. [12, 13]) it is an easy matter to check that for any vector $z \in \mathbf{R}^n$

$$\mathcal{P}_{\mathcal{R}(A), \mathcal{N}(A)} \, z = z - \left(\sum_{i=1}^{n} z_i \right) / n \ , \tag{3}$$

which shows how the projections needed in the *PCG* step can be performed.

In Tables 1-4 are gathered the results of our numerical experiments, including the (effective (only the smallest positive eigenvalue is discarded as far as perturbed methods are concerned)) spectral condition number, the number of *PCG* iterations needed to reduce an initial residual by a prescribed amount and, for $k = 3$ the distribution of the extremal eigenvalues for $h^{-1} = 192$. The computations were carried out in double precision Fortran on a *CDC 4680* computer.

The merits of our treatment of difficult problems with anisotropy are clearly displayed. Very often, an important reduction in the (effective) spectral condition number is observed, which reflects on the rate of convergence of the *PCG* method.

We end by stressing that our conclusions do not depend on the fact that the linear system to solve is singular [13]. For space limitations and since much is known about nonsingular block preconditionings, we have taken this opportunity to speak a bit on singular block methods.

region	p	q
Ω_1	1	1
Ω_2	1	10^k
Ω_3	10^k	1

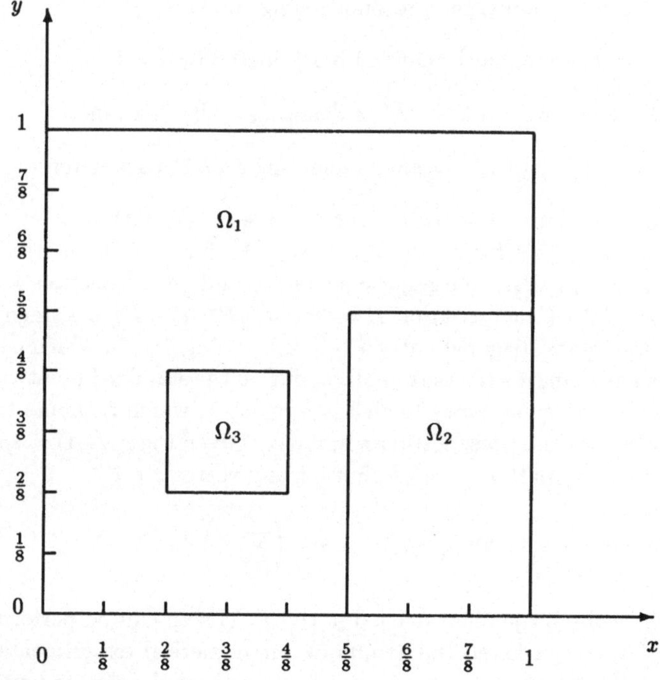

Figure 1: Specification of the coefficients of the partial differential equation.

Table 1: Spectral condition number of the preconditioned matrix B^+A associated with $MBILU$, $EMBILU(\alpha)$ and $DMBILU(\alpha)$ with $\alpha = 1/M$ for $k = 3$; exponent μ corresponding to the (estimated) asymptotic relationship $\kappa(B^+A) = Ch^{-\mu}$, C denoting a constant.

h^{-1} preconditioning	24	48	96	192	μ
$MBILU$	37.92	94.00	231.0	599.3	1.38
$DMBILU(\alpha)$ $s=1$	148.5	300.6	693.1	1659	1.26
$EMBILU(\alpha)$ $s=1$	6.96	12.78	27.31	61.83	1.18

Table 2: Distribution of extremal eigenvalues, spectral condition number κ and effective spectral condition number κ_{eff} for $k = 3$, $h^{-1} = 192$ and for $MBILU$, $DMBILU(\alpha)$ and $EMBILU(\alpha)$ with $\alpha = 1/M$.

preconditioning	smallest eigenvalues				largest eigenvalues				κ	κ_{eff}
MBILU	1.	1.	1.	1.	168	222	262	599	599	262
DMBILU(α) $s=1$	0.033	0.11	0.23	0.37	27	34	42	54	1659	478
EMBILU(α) $s=1$	0.90	0.96	0.97	0.98	41	44	50	56	62	58

Table 3: Number of PCG iterations to achieve $\|r^{(i)}\|_2 / \|r^{(0)}\|_2 \leq \epsilon$ for $k=3$ and $h^{-1}=192$.

ϵ preconditioning	10^{-3}	10^{-5}	10^{-7}	10^{-9}
MBILU	24	46	64	84
DMBILU(α) $s=1$	27	59	72	89
DMBILU(α) $s=\frac{1}{4}$	57	105	121	135
EMBILU$(\alpha)=1$	15	27	42	53
EMBILU$(\alpha)=\frac{1}{4}$	10	25	35	44

Table 4: Number of *PCG* iterations to achieve $\|r^{(i)}\|_2 / \|r^{(0)}\|_2 \le \epsilon$ for $h^{-1} = 192$ and $k = 1, 2$.

ϵ	10^{-3}	10^{-5}	10^{-7}	10^{-9}	10^{-3}	10^{-5}	10^{-7}	10^{-9}
	$k = 1$				$k = 2$			
MBILU	32	49	70	87	31	50	69	87
DMBILU(α)								
$s = 1$	20	33	44	56	24	37	49	61
$s = 1/4$	25	36	44	54	34	51	63	72
EMBILU(α)								
$s = 1$	23	36	48	60	20	34	46	58
$s = 1/4$	23	35	46	56	21	32	43	53

References

[1] O. AXELSSON, *On the eigenvalue distribution of relaxed incomplete factorization methods and the rate of convergence of conjugate gradient methods*, Technical Report, Departement of Mathematics, Catholic University, Nijmegen, The Netherlands, 1989.

[2] O. AXELSSON AND V. BARKER, *Finite Element Solution of Boundary Value Problems. Theory and Computation*, Academic Press, New York, 1984.

[3] O. AXELSSON AND G. LINDSKOG, *On the eigenvalue distribution of a class of preconditioning methods*, Numer. Math., 48 (1986), pp. 479–498.

[4] R. BEAUWENS, *Lower eigenvalue bounds for pencils of matrices*, Lin. Alg. Appl., 85 (1987), pp. 101–119.

[5] R. BEAUWENS AND M. BEN BOUZID, *Existence and conditioning properties of sparse approximate block factorizations*, SIAM J. Numer. Anal. , 25 (1988), pp. 941–956.

[6] A. BEN ISRAEL AND T.N.E. GREVILLE, *Generalized Inverses : Theory & Applications*, J. Wiley & Sons, New York, 1974.

[7] A. GREENBAUM, *Behavior of slightly perturbed lanczos and conjugate-gradient recurrences*, Lin. Alg. Appl., 113 (1989), pp. 7–63.

[8] M.M. MAGOLU, *Conditioning analysis of sparse block approximate factorizations*, Appl. Numer. Math., 8 (1991), pp. 25-42.

[9] M.M. MAGOLU, *Analytical bounds for block approximate factorization methods*, (1992), Lin. Alg. Appl. To appear.

[10] M.M. MAGOLU, *Modified block-approximate factorization strategies*, Numer. Math. 61 (1992), pp. 91-110.

[11] M.M. MAGOLU, *Lower eigenvalue bounds for singular pencils of matrices*, (1992), J. Comput. Appl. Math. To appear.

[12] M.M. MAGOLU, *Sparse block approximate factorizations for singular problems*, in Iterative Methods in Linear Algebra, R. Beauwens and P. de Groen, eds., North-Holland, 1992.

[13] M.M. MAGOLU, *Sparse approximate block factorizations for solving symmetric positive (semi)definite linear systems*, thèse de Doctorat, Université Libre de Bruxelles, Brussels, Belgium, 1992.

[14] M.M. MAGOLU AND Y. NOTAY, *On the conditioning analysis of block approximate factorization methods*, Lin. Alg. Appl., 154-156 (1991), pp. 583–599.

[15] Y. NOTAY, *Polynomial acceleration of iterative schemes associated with subproper splittings*, J. Comput. Appl. Math., 24 (1988), pp. 153–167.

[16] Y. NOTAY, *Résolution iterative de systèmes linéaires par factorisations approchées*, thèse de Doctorat, Université Libre de Bruxelles, Brussels, Belgium, 1991.

[17] Y. NOTAY, *Conditioning analysis of modified block incomplete factorizations*, Lin. Alg. Appl., 154-156 (1991), pp. 711–722.

[18] A. VAN DER SLUIS, *The convergence behaviour of conjugate gradients and ritz values in various circumstances*, in Iterative Methods in Linear Algebra, R. Beauwens and P. de Groen, eds., North-Holland, 1992.

[19] H.A. VAN DER VORST, *The convergence behaviour of preconditioned CG and CG-S*, in Preconditioned Conjugate Gradient Methods, O. Axelsson and L. Kolotilina, eds., Lectures Notes in Mathematics No. 1457, Springer-Verlag, 1990, pp. 126–136.

Preconditioning of Blockstructured Linear Systems with Block ILU on Parallel Computers

Volker Mehrmann, Fakultät für Mathematik, Universität Bielefeld,
Postfach 8640, D-4800 Bielefeld 1, FRG, Tel.: (0521) 106-4798,
email: mehrmann@math1.mathematik.uni-bielefeld.de

Abstract

We discuss block incomplete factorizations for large sparse block tridiagonal matrices arising for example in the numerical solution of partial differential equations. We describe a procedure to partition the matrices into subsystems that can be easily, efficiently and in a numerically stable way solved on a parallel computer. The partitioning is based on low rank modifications which generalize the approaches used in Divide and Conquer methods. The effect of these preconditioners are demonstrated for two model problems.

1　Introduction

We discuss a purely algebraic approach to compute preconditioners in iterative methods for the solution of block tridiagonal systems $TY = C$ on a parallel computer. We consider systems of the form

$$
\begin{bmatrix}
T_{11} & T_{12} & & & \\
T_{21} & T_{22} & T_{23} & & \\
& \ddots & \ddots & \ddots & \\
& & \ddots & \ddots & T_{m-1,m} \\
& & & T_{m,m-1} & T_{m,m}
\end{bmatrix}
\begin{bmatrix}
Y_1 \\
Y_2 \\
\vdots \\
Y_{m-1} \\
Y_m
\end{bmatrix}
=
\begin{bmatrix}
C_1 \\
C_2 \\
\vdots \\
C_{m-1} \\
C_m
\end{bmatrix},
\tag{1}
$$

with square blocks T_{ii} of sizes $k_i \times k_i$, $i = 1, \ldots, m$ and (possibly several) right sides C_i of sizes $k_i \times r$, $i = 1, \ldots, m$. The whole system is of size $s = \sum_{i=1}^{m} k_i$. The preconditioners are constructed in such a way to avoid communication between different processors of a multiprocessor system, while at the same time obtaining a good preconditioner. We can formulate the problem as follows:

Problem 1 Find a matrix $\tilde{T}$ that satisfies the following **principle goals for the preconditioning:**

- (a) the condition number of $T_p := \tilde{T}^{-1} T$ is as small as possible,

- (b) linear systems with system matrix $\tilde{T}$ are easy and inexpensive to solve,

- (c) linear systems with system matrix $\tilde{T}$ are easy to solve on a parallel computer,

- (d) the numerical solution of linear systems with system matrix $\tilde{T}$ is well conditioned, i.e. $\operatorname{cond}(\tilde{T})$ is as small as possible,

- (e) the numerical method for systems with system matrix $\tilde{T}$ is stable,

- (f) the method to compute $\tilde{T}$ is easily parallelizable and numerically stable.

There is usually a trade-off between different aspects of (a),...,(f)!

A well established tool to perform such a preconditioning in the case of matrices arising in the numerical solution of partial differential equations is the incomplete LU factorization (ILU) or in the case discussed here better the incomplete block LU factorization. By a block ILU factorization, we will understand the following:

Given a block matrix A and a block graph G, express A as

$$A = LU + N, \tag{2}$$

where the block graph of L and the block graph of U are subgraphs of G and L, U^T are lower triangular. Here the block graph of a block matrix $A = [A_{ij}]$ with $m \times m$ blocks is the directed graph with m vertices and edges from i to j if $A_{ij} \neq 0$.

Often in applications G is chosen to be the block graph of A, so L, U have block entries at most, where A has a block entry. In the case of a block tridiagonal matrix, however, this would yield a complete block LU factorization if we choose G to be the block graph of A.

The block LU factorization, on the other hand, is difficult to parallelize, since the process is highly recursive. Thus, in order to obtain a method that parallelizes well we have to take a graph G for the block ILU that is not as strongly connected as the graph of the block tridiagonal matrix, which is just a path.

In this paper we discuss an approach that sacrifices a bit of the property (a) in Problem 1 while improving on parts (b), (c), (d), (e), (f). A similar approach can also be applied to general sparse block structured matrices but for the sake of avoiding complicated notation we will restrict ourselves to block tridiagonal systems.

2   A parallel approach

The first step of our approach is to repartition the system into the new system $SX = B$ of $p \leq m$ blocks (p is numbers of available processors).

$$\begin{bmatrix} S_{11} & S_{12} & & & \\ S_{21} & S_{22} & \ddots & & \\ & \ddots & \ddots & \ddots & \\ & & \ddots & \ddots & S_{p-1,p} \\ & & & S_{p,p-1} & S_{p,p} \end{bmatrix} \begin{bmatrix} X_1 \\ X_2 \\ \vdots \\ X_p \end{bmatrix} = \begin{bmatrix} B_1 \\ B_2 \\ \vdots \\ B_p \end{bmatrix}, \tag{3}$$

with S_{ij} of size $d_i \times d_j$, where the m blocks of T are partitioned, such that each of the p diagonal blocks of S is quadratic and has (if possible) equally many rows.

Observe that for $p < m$ we have

$$S_{j,j+1} = \begin{bmatrix} 0 & \cdots & \cdots & 0 \\ \vdots & & & \vdots \\ 0 & \cdots & \cdots & 0 \\ T_{i_j,i_{j+1}} & 0 & \cdots & 0 \end{bmatrix} = T_{i_j,i_{j+1}} E_j F_j^T, \tag{4}$$

where

$$E_j := \begin{bmatrix} 0 \\ \vdots \\ 0 \\ I \end{bmatrix}, F_j := \begin{bmatrix} I \\ 0 \\ \vdots \\ 0 \end{bmatrix}.$$

Analogously
$$S_{j+1,j} = T_{i_{j+1},i_j} F_j E_j^T.$$

In applications coming from partial differential equations, this is often easy to achieve, by making the grid size apropriate. In other applications or in case of unstructered grids we may use a partitioning that is oriented at the block structure of T in such a way that the maximal size of a block in S is minimal and the blocks are of almost equal size. For a discussion of this combinatorial optimization problem see [8]. We then rewrite (3) as

$$(\tilde{S} + W)X := \left(\begin{bmatrix} \tilde{S}_{11} & & \\ & \ddots & \\ & & \tilde{S}_{pp} \end{bmatrix} + \sum_{j=1}^{p-1} W_j \right) \begin{bmatrix} X_1 \\ X_2 \\ \vdots \\ X_p \end{bmatrix} = \begin{bmatrix} B_1 \\ B_2 \\ \vdots \\ B_p \end{bmatrix}, \tag{5}$$

where

$$W_j = \mathcal{E}_j \begin{bmatrix} U_{jj} & U_{j,j+1} \\ V_{j+1,j} & V_{j+1,j+1} \end{bmatrix} \mathcal{E}_j^T, \tag{6}$$

with

$$\mathcal{E}_j := \begin{bmatrix} E_j & 0 \\ 0 & F_j \end{bmatrix} \tag{7}$$

and where

$$U_{j,j+1} = T_{i_j,i_j+1}, \quad V_{j+1,j} = T_{i_j+1,i_j}, \tag{8}$$

and $U_{j,j}, V_{j+1,j+1}$ are to be chosen freely, so that all systems $\tilde{S}_{jj}$ are solvable.

We then precondition the system with $\tilde{S}^{-1}$ and obtain the preconditioned system

$$(I - \tilde{W})X := (I - \tilde{S}^{-1}W)X = \tilde{S}^{-1}B, \tag{9}$$

which due to the fact that $\tilde{W}$ is of small rank, is a small rank modification of the identity matrix. Thus it is well known that conjugate gradient type methods, see [5], are known to converge theoretically in $\mathrm{rank}(\tilde{W})$ steps.

We now discuss two possibilities to choose the blocks U_{ii}, V_{ii}. Note that the matrices $U_{j,j+1}$, $V_{j+1,j}$ are already fixed in order to achieve zero blocks in the off–diagonal positions. The first and simplest choice that one could take is to choose the diagonal blocks to be zero, i.e.

$$U_{jj} = 0, \quad V_{j+1,j+1} = 0; \quad j = 1,\ldots,p-1. \tag{10}$$

This choice is in the context of direct methods for the solution of (3) also used in other divide and conquer approaches, see [2, 7]. One obtains

$$\tilde{S} = \mathrm{diag}(S_{11},\ldots,S_{pp}) \tag{11}$$

and it is clear that linear systems with $\tilde{S}$ are easy to solve on parallel computer with p processors, if each of the processors gets one of the blocks S_{jj}. If the systems with these blocks are solved via an LU decomposition $S_{jj} = L_{jj}U_{jj}$, then this gives a block ILU factorization, where the specified block graph is that of the matrix $\tilde{S}$.

Observe that this choice of $\tilde{S}$ is not the one which yields a W of minimal rank. If all the matrices $T_{k_j,k_j+1}, T_{k_j+1,k_j}$ have full rank, then this is the maximal rank decoupling among all choices of the given form.

The critical point with this approach is the numerical stabilty. If the matrix T is symmetric positive definite or an M-matrix then by this approach we will also yield a matrix with the same properties and it is well known that then the block LU factorization without pivoting is numerically stable e.g. [1, 6, 10]. In indefinite or unsymmetric cases, however, this approach maybe be unstable since it may even happen that $\tilde{S}$ is singular. Consider the following example:

Example 1 In the splitting

$$T = \tilde{T} + W := \begin{bmatrix} 1 & 1 & 0 & 0 \\ 1 & 1 & 0 & 0 \\ 0 & 0 & 1 & 1 \\ 0 & 0 & 1 & 1 \end{bmatrix} + \begin{bmatrix} 0 & 0 & 0 & 0 \\ 0 & 0 & 1 & 0 \\ 0 & 1 & 0 & 0 \\ 0 & 0 & 0 & 0 \end{bmatrix} \tag{12}$$

we have T nonsingular indefinite, but $\tilde{T}$ is singular.

Another approach is the idea of minimal rank decoupling which in the case of a complete block LU factorization is first discussed in [8]. For this decoupling we require that in (6) we have

$$\begin{bmatrix} U_{jj} & U_{j,j+1} \\ V_{j+1,j} & V_{j+1,j+1} \end{bmatrix} = \begin{bmatrix} M_{jj} \\ N_{j+1,j} \end{bmatrix} \begin{bmatrix} \hat{M}_{jj}^T & \hat{N}_{j+1,j}^T \end{bmatrix}, \tag{13}$$

i.e. the block 2×2 matrices are written as outer products, with the constraint that

$$N_{j+1,j}\,\hat{M}_{jj}^T = V_{j+1,j}, \quad M_{jj}\,\hat{N}_{j+1,j}^T = U_{j,j+1} \tag{14}$$

are given.

In the case of a scalar tridiagonal matrix this leads to rank 1 matrices in (13) and clearly if $V_{j+1,j}$, $U_{j,j+1}$ are full rank, then this is a matrix of smallest rank satisfying the constraints.

There is still freedom in the choice of the diagonal blocks subject to (14). If our original system (1) was symmetric positive definite, then we have that in (13) $U_{j,j+1} = V_{j+1,j}^T$ and we can choose the factorizations in (13) such that

$$M_{jj} = \hat{M}_{jj}, \quad N_{j+1,j} = \hat{N}_{j+1,j}. \tag{15}$$

or

$$M_{jj} = -\hat{M}_{jj}, \quad N_{j+1,j} = -\hat{N}_{j+1,j}. \tag{16}$$

Possible choices for this approach and their properties have been discussed in [8, 9]. In general we can formulate the problem of choosing the modification as the following **open problem**:

Problem 2 Given a block tridiagonal system of the form (1) partitioned in the form (3). Choose

$$M_{jj}, N_{j+1,j}, \hat{M}_{jj}, \hat{N}_{j+1,j}; \quad j = 1, \ldots, p-1 \tag{17}$$

such that

$$N_{j+1,j}\hat{M}_{jj}^T = T_{i_j+1,i_j}, \quad M_{jj}\hat{N}_{j+1,j}^T = T_{i_j,i_j+1} \tag{18}$$

and such that the matrices $\tilde{S}$ and $\tilde{S}^{-1}S$ with these choices are optimally conditioned.

Observe that this is not a standard type block ILU factorization.

In order to see what the discussed approach does consider the case that the linear system S is obtained from the finite difference equation of the Poisson equation

$$\begin{aligned} -\Delta u &= f \quad \text{in} \quad \Omega = (0,a) \times (0,b) \\ u &= g \quad \text{on} \quad \delta\Omega, \quad \text{where} \quad \Delta = \tfrac{\partial^2}{\partial x^2} + \tfrac{\partial^2}{\partial y^2}. \end{aligned} \tag{19}$$

with the standard 5–point difference scheme with $n * m$ interior grid points in a regular mesh of meshsize h in Ω, e.g. [6, 10]. For (19) we obtain a block matrix

$$
T = \begin{bmatrix}
T_{11} & T_{12} & & & \\
T_{21} & T_{22} & \ddots & & \\
& \ddots & \ddots & & \ddots \\
& & \ddots & \ddots & T_{m-1,m} \\
& & & T_{m,m-1} & T_{m,m}
\end{bmatrix}
\tag{20}
$$

with

$$
T_{ii} = \begin{bmatrix}
4 & -1 & & & \\
-1 & 4 & -1 & & \\
& \ddots & \ddots & \ddots & \\
& & & & -1 \\
& & & -1 & 4
\end{bmatrix}
\quad \text{of size} \quad n \times n
\tag{21}
$$

and $T_{i,i+1} = T_{i+1,i} = -I_n$.

For the two small rank modification problems the partitioning can be interpreted as a domain decomposition approach, e.g. [6], into subdomains which have no common points and appropriately chosen boundary conditions for the points next to an internal boundary. For such points we obtain for the maximal rank decoupling the stencil

$$
\boxed{\begin{array}{ccccc}
 & & -1 & & \\
 & & \updownarrow & & \\
-1 & \leftrightarrow & 4 & \leftrightarrow & -1
\end{array}}
\tag{22}
$$

and for the minimal rank decoupling the stencil

$$
\boxed{\begin{array}{ccccc}
 & & -1 & & \\
 & & \updownarrow & & \\
-1 & \leftrightarrow & 5 & \leftrightarrow & -1
\end{array}}
\tag{23}
$$

3 Comparison of the methods

We now compare the two preconditioning approaches described above.

Let $csmax$, $csmin$ denote the condition numbers of the matrices $\tilde{S}$ in the two approaches and let $cpmax$ and $cpmin$ be condition numbers for the preconditioned matrices $\tilde{S}^{-1}S$, respectively.

We discuss two model problems. The first is the Poisson equation as discussed above and the second is a simplification of a system that is obtained from a flux vector splitting method for the Euler equations of fluid dynamics, e.g. [3, 4]. The linear system has a system matrix

$$
T = \begin{bmatrix}
T_1 & T_2 & & & \\
T_3 & T_1 & \ddots & & \\
& \ddots & \ddots & \ddots & \\
& & \ddots & \ddots & T_2 \\
& & & T_3 & T_1
\end{bmatrix}
\tag{24}
$$

with n block rows, where

$$T_1 = \begin{bmatrix} C & -A^- & & & \\ -A^+ & C & \ddots & & \\ & \ddots & \ddots & \ddots & \\ & & \ddots & \ddots & -A^- \\ & & & -A^+ & C \end{bmatrix}, \tag{25}$$

$T_2 = \mathrm{diag}(-B^-,\ldots,-B^-)$, $T_3 = \mathrm{diag}(-B^+,\ldots,-B^+)$ each with m block rows, and A^+, A^-, B^+, B^- are negative semidefinite matrices obtained by splitting symmetric indefinite matrices A, B as $A = A^+ - A^-, B = B^+ - B^-$, while $C = A^+ + A^- + B^+ + B^-$. In our test example we used

$$A = \begin{bmatrix} .438 & .981 & .713 & .687 \\ .982 & .760 & .573 & 1.21 \\ .713 & .573 & 1.06 & .738 \\ ..687 & 1.21 & .738 & .835 \end{bmatrix}, B = \begin{bmatrix} 1.37 & 1.12 & 1.63 & .894 \\ 1.12 & .184 & 1.56 & 1.15 \\ 1.63 & 1.56 & 1.52 & .591 \\ .894 & 1.15 & .591 & 1.27 \end{bmatrix}, \tag{26}$$

which are randomly chosen symmetric matrices each with 2 positive and 2 negative eigenvalues. Observe that due to the special form $A^+ A^- = A^- A^+ = 0$ and the off diagonal blocks are all of rank 2 and it follows in this case that the matrices $W, \tilde{W}$ have the same rank for both approaches.

For the Poisson equation with a grid of 8×64 interior points, i.e. $n * m = 512$, the condition number of the unpartitioned matrix is $\mathrm{cond}_2(T) = 6.41E1$.

For different p we obtain here:

$$\begin{array}{c c c c c}
p & \text{csmax} & \text{csmin} & \text{cpmax} & \text{cpmin} \\
2 & 6.07E1 & 6.07E1 & 8.05 & 9.95 \\
4 & 5.07E1 & 5.03E1 & 8.20 & 10.2 \\
8 & 3.22E1 & 3.08E1 & 9.80 & 1.26E1 \\
16 & 1.49E1 & 1.34E1 & 1.31E1 & 1.58E1
\end{array} \tag{27}$$

For the other problem the condition number of the unpartitioned matrix is $\mathrm{cond}_2(T) = 1.99E2$. For different p we obtain here:

$$\begin{array}{c c c c c}
p & \text{csmax} & \text{csmin} & \text{cpmax} & \text{cpmin} \\
2 & 1.20E2 & 1.79E2 & 2.58E1 & 2.66E1 \\
4 & 7.61E1 & 1.67E2 & 3.10E1 & 3.16E1 \\
8 & 6.47E1 & 1.48E2 & 3.63E1 & 3.58E1 \\
16 & 5.18E1 & 1.31E2 & 4.28E1 & 4.24E1
\end{array} \tag{28}$$

To see the effect of preconditioning for the second example we give the convergence history for the BCG,CGS and BCGSTAB conjugate gradient type methods, e.g. [5] for both approaches with $p = 4, 8$. Observe that in both approaches the rank of the preconditioned matrix is $k = 16, 32$ respectively. Thus in exact arithmetic each of the three converges in $k = 16, 32$ steps. This property can be clearly observed for the stabilized BCGSTAB method.

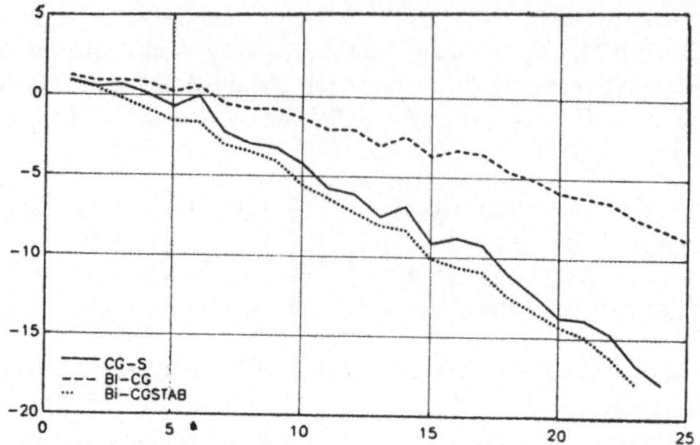

Figure 1: Number of iterations versus log_{10} of the norm of the residual for $p = 4$.

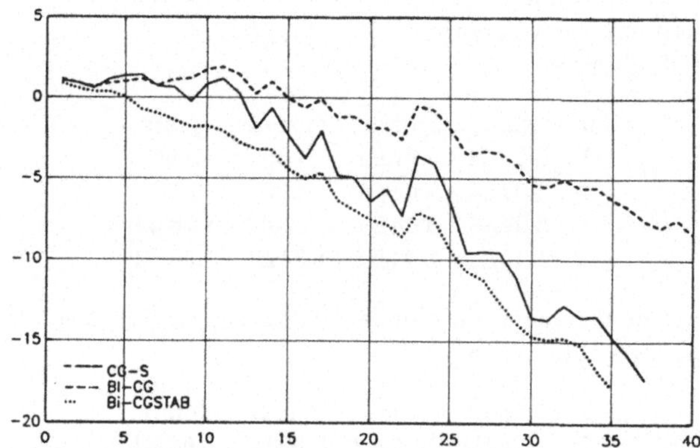

Figure 2: Number of iterations versus log_{10} of the norm of the residual for $p = 8$.

4   Conclusion

We have described two methods for the preconditioning of block tridiagonal linear systems on a parallel computer via decoupling by low rank modifications.

For the minimal rank modification we have the following properties:

- (a) cond $(\tilde{S}^{-1}S)$ is not as small as possible but improved,

- (b) systems with $\tilde{S}$ are easy and inexpensive to solve,

- (c) systems with $\tilde{S}$ are easy to parallelize,

- (d) systems with $\tilde{S}$ can be made well conditioned,

- (e) stable numerical methods can be chosen,

- (f) $\tilde{S}$ is easily and stably obtained on a parallel machine.

Furthermore we have that $\tilde{S}^{-1}S$ is a small rank (k) modification of the identity, i.e. CG type methods converges theoretically in k steps.

For the maximal rank modification we have similar properties but only for symmetric positive definite M-matrices or matrices with similar properties, the numerical stability is garanteed.

References

[1] A. BERMAN AND R. PLEMMONS, *Nonnegative Matrices in the Mathematical Sciences*, Academic Press, New York, 1979.

[2] S. BONDELI, *Divide and Conquer: Parallele Algorithmen zur Loesung tridiagonaler Gleichungssysteme, Dissertation, ETH Zuerich*, Verlag der Fachvereine, Zuerich, 1991.

[3] E. DICK AND J. LINDEN, *A multigrid flux-difference splitting method for steady incompressible Navier-Stokes equations*, Proceedings of the GAMM Conference on Numerical Methods in Fluid Mechanics, Delft, (1989).

[4] L. ELSNER AND V. MEHRMANN, *Convergence of block iterative methods for linear systems arising in the numerical solution of Euler equations*, Numer. Math., 59 (1991), pp. 541–560.

[5] R. FREUND, G. H. GOLUB, AND N. NACHTIGAL, *Iterative solution of linear systems*, Tech. Report na-91-05, Computer Science Dept., Stanford University, Stanford, Ca. 94305, 1991.

[6] W. HACKBUSCH, *Iterative Lösung großer schwachbesetzter Gleichungssysteme*, Teubner Verlag, Stuttgart, 1991.

[7] A. KRECHEL, H. PLUM, AND K. STUEBEN, *Solving tridiagonal linear systems in parallel on local memory mimd machines*, tech. report, Gesellschaft f. Mathematik und Datenverarbeitung, St. Augustin (FRG), 1989.

[8] V. MEHRMANN, *Divide and conquer methods for block tridiagonal systems*, tech. report, Institut für Geometrie und Praktische Mathematik, RWTH Aachen, Aachen FRG, 1991.

[9] V. MEHRMANN, *Divide and conquer methods for tridiagonal linear systems*, Notes on Numerical Fluid Mechanics (Parallel Algorithms for Partial Differential Equations), (Proceedings of the Sixth GAMM-Seminar, Kiel, January 19-21, 1990 Wolfgang Hackbusch, Editor), (1991).

[10] R. VARGA, *Matrix Iterative Analysis*, Prentice-Hall, Englewood Cliffs, N.J., 1962.

Quadrant Tridiagonal Partitioning and Preconditioned

Conjugate Gradient Method for Solving Elliptic Problems

by

G. Molnárka and S. Szabó
Department of Numerical Analysis
Eötvös University,Budapest

Introduction

Quadarant diagonal partitioning (QDP) method is an iterative method for solving linear system proposed by D. J. Evans et al. (see [1],[2],[3]) appropriate for paralell implementation. Here we show that in case of second order elliptic problem, the classical, color and multicolor ordering technique [4] are applicable for further paralellization of QDP iterative methods. Moreover we give a direct algorithm with arithmetical complexity O(n) for solving a special quadrant tridiagonal matrix and using these results we propose a quadrant tridiagonal precondititoned (QTP) conjugate gradient method for solving second order elliptic problems. The numerical experiments presented here shows that this method in several cases are more effective than other PCG algorithm.

1. Quadrant Tridiagonal Partitioning and elliptic problems

Iterative methods similar to the point Jacobi and successive over relaxation like to the LU partitioning may be devised by quadrant diagonal partitioning: $A = X - W - Z$ too, where X has nonzeros only on the main and cross diagonals, $-W$ and $-Z$ have zeros on the the main and cross diagonals and nonzeros on position given by the following formula:

$$(X)_{ij} = \begin{cases} 0 & if \quad i \neq j \quad and \quad i \neq n - j + 1 \\ a_{i,j} & otherwise \end{cases}$$

$$(-W)_{ij} = \begin{cases} 0 & if \quad i \leq j \leq n - i + 1 \quad or \quad n - i + 1 \leq j \leq i \\ a_{ij} & otherwise \end{cases} \tag{1}$$

$$(-Z)_{ij} = \begin{cases} 0 & if \quad j \leq i \leq n - j + 1 \quad or \quad n - j + 1 \leq i \leq j \\ a_{ij} & otherwise \end{cases}$$

Using such partitioning the successive overrelaxation QDP method is as follows:

$$(X - \omega W)x^{(p+1)} = (\omega Z + (1 - \omega)X)x^{(p)} + \omega b \tag{2}$$

where ω is the relaxation factor. Such kind of iterative scheme is important because it can be regarded as (2x2) block-iterative methods, and lead naturally to paralellism [3]. The successive QDP method converges if A is irreducible and posseses weak diagonal dominance, or if A is real and positive definite for $0 < \omega < 2$

Let us regard the two dimensional elliptic problem on rectangular domain. As a simple representative let us examine the following problem:

$$\Delta u = f(x), \quad x \in \Omega, \quad u \mid_{\partial\Omega} = 0 \tag{3}$$

Now we should show,that using the classical five point finite difference stencil and a special ordering of grid points the resulting matrix approximating the original problem (3) will have such structure which is appropriate form for the QDP process. The following simple example shows one of such orderings.

Two neighbouring grid points we regard as one grid point omitting edge between them and these combined grid points we order in line (or row) continuous manner. Than the number of order if combined grid points numbered by p one split up into two grid points with order p and n-p+1, where n is the total number of grid points of lattice. This ordering process and the structure of resulting matrix can be illustrated by fig. 1a. and 1b.

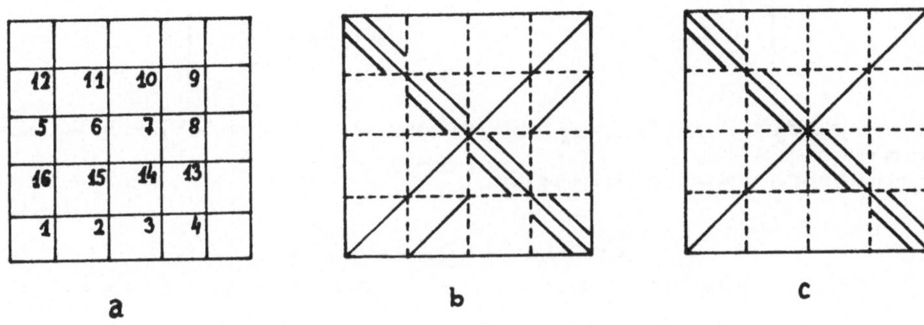

Fig. 1. The matrix structures and ordering

The ordering proposed here can be generalised by combining with different coloring techniques. For examle in case of five point difference stencil a two color ordering of "combined" grid points gives more possibilities of paralellisation, while for 9 point finite difference stars a four colour ordering [5] of "combined" grid points and than an above given splitting of combined grid points gives a matrix with structure useful for different paralellisation on a linear two dimensional lattice network of processors. These and further examples can be verified easily by calculating these structure.

The results sketched above raise the following question: could we use this matrix form for construction of effective iteration process and/or preconditioners that can be solved in parallel. If from the matrix illustrated in Fig.1b. we omit the nonzeros from the cross diagonal below the main cross diagonal Fig.1c, we get n such subproblems which can be solved in parallel furthermore in the next point we shall show, that the matrix of one subproblem can be solved directly in O(n) steps where $n = 2\sqrt{N}$, where N is the total number of grid points. Such matrices in the follwing we call quadrant tridiagonal (QT) matrices.

2. Direct algorithm for quadrant tridiagonal matrices

Our problem is to solve a linear system of equations:

$$Ax = r \tag{4}$$

where A has special structure explained above, namely in general the $(n \times n)$ matrix A of a subproblem has the following quadrant tridiagonal form:
let $n = 2k$,where k is integer and

$$
\begin{bmatrix}
a_1 & b_1 & & & & & & & d_1 \\
c_2 & a_2 & b_2 & & & & & d_2 & \\
& \cdot & \cdot & \cdot & & & \cdot & & \\
& & \cdot & \cdot & \cdot & \cdot & & & \\
& & & c_{\frac{n}{2}} & a_{\frac{n}{2}} & b_{\frac{n}{2}} & & & \\
& & & & c_{\frac{n}{2}+1} & a_{\frac{n}{2}+1} & b_{\frac{n}{2}+1} & & \\
& & & \cdot & & \cdot & \cdot & \cdot & \\
& & \cdot & & & & \cdot & \cdot & \cdot \\
& d_{n-1} & & & & & c_{n-1} & a_{n-1} & b_{n-1} \\
d_n & & & & & & & c_n & a_n
\end{bmatrix}
= A. \qquad (5)
$$

It is clear, that $d_{\frac{n}{2}} = b_{\frac{n}{2}}$ and $d_{\frac{n}{2}+1} = c_{\frac{n}{2}+1}$ The (4) system of equation can be writen by the folowing pairs of equations:
Let suppose that $c_1 = b_n = 0$, so (4) means that:

$$
c_i x_{i-1} + a_i x_i + b_i x_{i+1} + d_i x_{n-i+1} = r_i,
$$
$$
d_{n-i+1} x_i + c_{n-i+1} x_{n-i} + a_{n-i+1} x_{n-i+1} + b_{n-i+1} x_{n-i+2} = r_{n-i+1}, \qquad 1 \le i \le k \qquad (6)
$$

We would like to solve this system by recursion so we suppose,that the following relations hold:

$$
x_i = \alpha_{i+1} x_{i+1} + \beta_{n-i} x_{n-i} + \epsilon_i,
$$
$$
x_{n-i+1} = \gamma_{i+1} x_{i+1} + \delta_{n-i} x_{n-i} + \epsilon_{n-i+1}, \qquad i = 1, 2, ..., k, \qquad (7)
$$

where $\alpha_{i+1}, \beta_{n-i}, \gamma_{i+1}, \delta_{n-i}, \epsilon_i, \epsilon_{n-i+1}, i = 1, 2, ..., k$ are undetermined coefficients. The value of these undetermined coefficients we get by substituting (7) into (6). In the calculations it is useful to separate the case i=1 and other i values so the results are as follows:
for i=1,

$$
\alpha_2 = -\frac{1}{detA_1} b_1 a_n, \qquad\qquad \beta_{n-1} = \frac{1}{detA_1} d_1 c_n,
$$

$$
\gamma_2 = \frac{1}{detA_1} d_n b_1, \qquad\qquad \delta_{n-1} = -\frac{1}{detA_1} a_1 c_n,
$$

$$
\epsilon_1 = \frac{1}{detA_1}(a_n r_1 - d_1 r_n), \qquad\qquad \epsilon_n = \frac{1}{detA_1}(a_1 r_n - d_n r_1),
$$

where $detA_1 = a_1 a_n - d_1 d_n$,
for i=2,3,...,k-1,

$$
\alpha_{i+1} = -\frac{1}{detA_i}(a_{n-i+1} + b_{n-i+1}\delta_{n-i+1}) b_i,
$$

$$
\beta_{n-i} = \frac{1}{detA_i}(c_i \beta_{n-i+1} + d_i) c_{n-i+1},
$$

$$
\gamma_{i+1} = \frac{1}{detA_i}(d_{n-i+1} + b_{n-i+1}\delta_{n-i+1}) b_i,
$$

$$
\delta_{n-i} = -\frac{1}{detA_i}(c_i \alpha_i + a_i) c_{n-i+1},
$$

$$
\epsilon_i = \frac{1}{detA_i}[(a_{n-i+1} + b_{n-i+1}\delta_{n-i+1})(r_i - c_i \epsilon_{i-1})) + (c_i \beta_{n-i+1} + d_i)(b_{n-i+1}\epsilon_{n-i+2} - r_{n-i+1})],
$$

$$\epsilon_{n-i+1} = \frac{1}{det\,A_i}[(d_{n-i+1} + b_{n-i+1}\gamma_i)(c_i\epsilon_{i-1} - r_i) + (c_i\alpha_i + a_i)(r_{n-i+1} - b_{n-i+1}\epsilon_{n-i+2})],$$

where

$$det\,A_i = (c_i\alpha_i + a_i)(a_{n-i+1} + b_{n-i+1}\delta_{n-i+1}) - (c_i\beta_{n-i+1} + d_i)(d_{n-i+1} + b_{n-i+1}\gamma_i).$$

Knowing the values $\alpha_{k+1}, \beta_k, \epsilon_k, \gamma_{k+1}, \delta_k, \epsilon_{k+1}$ the values x_k and x_{k+1} can be expressed from (7) directly. They are as follows:

$$x_k = \frac{1}{det\,B}[(1 - \gamma_{k+1})\epsilon_k + \alpha_{k+1}\epsilon_{k+1}],$$

$$x_{k+1} = \frac{1}{det\,B}[\delta_k\epsilon_k + (1 - \beta_k)\epsilon_{k+1}],$$

where

$$det\,B = (1 - \beta_k)(1 - \gamma_k) - \alpha_{k+1}\delta_k.$$

The values of x_i, x_{n-i+1} $\quad i = k - 1, k - 2, ..., 1$ can be determined by substitution and recursion using (7).

The required number of multiplications to realise this algorithm is $32k = 16n$ which roughly is only two times more than the number of multiplications required the well known so called reduced Gaussian elimination algorithm for tridiagonal matrices, but in this algorithm there are two independent recursions which can be performed in parallel.

The numerical stability properties of this algorithm theoritically we have not analysed yet. Because this algorithm is a short version of Gaussian elimination without pivoting its numerical stability properties must bew close to that one of Gaussian elimination. Really the numerical experiments show that the weak diagonal dominance of A assures the numerical stability of the algorithm proposed above.

This algorithm can be used for constructing different paralell iterative schemes to solve second order elliptic problems. Here we use this algorithm for constructing preconditions for conjugate gradient algorithm.

3. Quadrant tridiagonal matrix as preconditioner

One of the most succesful method for the solution of positive definite symmetric systems arising from the discretization of partial differential equations is the preconditioned conjugate gradient (PCG) method. We shall consider here the numerical solution of algebraic systems

$$Ax = b \qquad x, b \in R^N,$$

where the matrix A arise from the discretization of problem (3) The PCG method for solving this system based on the cojugate gradient solution of equivalent system

$$C^{-1}Ax = C^{-1}b,$$

where C is often called the preconditioning matrix. The solution of linear system with C must be relatively inexpensive. One can show that if one can construct a peconditioner such that the spectrum of the resulting matrix $C^{-1}A$ is contained in an ellipse which is away from the origin and has a fixed size independent of the problem size, then we get a convergence rate independent of the number of unknowns. Thus the task of deriving efficient preconditioning matrices is essential. Many kinds of both point and block preconditioners are possible. The most important methods of precondition construction one can find in ([7] and [8]). Our proposition for preconditioners here has some formal character, because we have seen only the special structure of matrix A and possibility of parallelisation of the preconditioners and

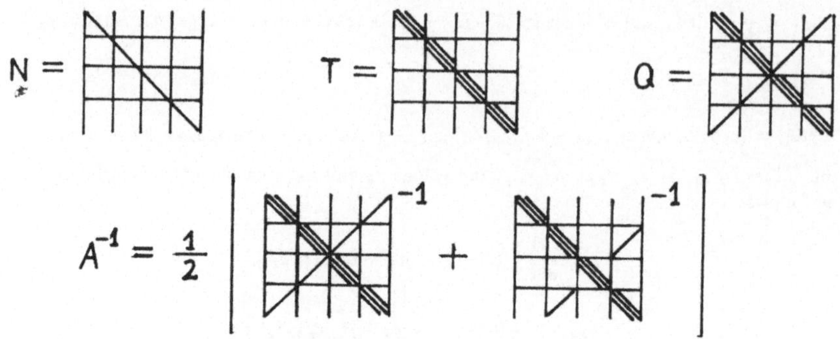

Fig.2 The different form of preconditioners

and we haven't analized yet the spectral properties of the preconditioners. The form of proposed and by numerical experiments analized preconditioners are shown on Fig. 2.

4. Numerical results

We have performed some numerical experiments in order to obtain some results to characterize the above proposed algorithms . The test program was written in C language and was running on an IBM 80386-387 computer. At the first we can notice, that the pure algorithm without preconditioning was able to solve the problems by several times less iteration, than the theoretical estimation therefore we haven't done any theoretical estimation.

In the problems 1-3 was the linear systems arising from discretization of elliptic problems. The solution x was given, and the vector $b = Ax$ was calculated. 3-D outlining of the solution of problems 1-3. can be seen on fig.3. The "special matrix" is a discretization of an elliptic problem with discontinuous coefficients.

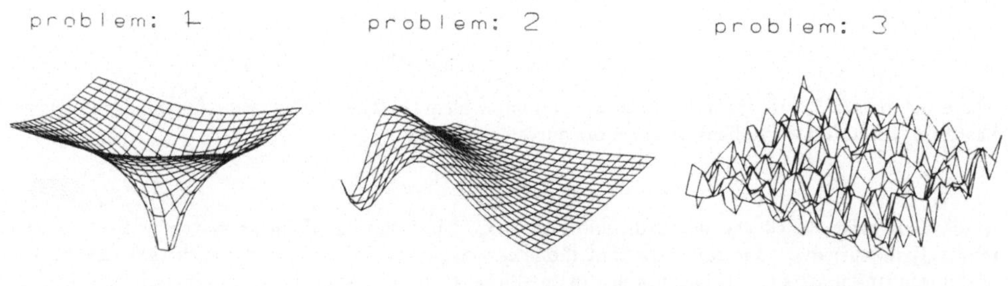

Fig. 3.

To be able to compare the accuracy we defined an error as follows:

$$error = \frac{\|x^{(p)} - x\|^2}{\|x\|^2}$$

About the preconditioners: We considered the well-known tridiagonal matrix preconditioner (on the fig. denoted by T), quadrant tridiagonal (on fig. denoted by Q) and a composed preconditioner (denoted by A). A symple CG method as a standard to compare with is denoted by N. This preconditioners have simple structure, and showed to be flexible. The effect given by the preconditioners slightly depends on the problem to be solved, see the fig.4 and on fig.5.

While A contains several small blocks, and a pair of them is chosen to be processed by the quadrant tridiagonal solver. These processes are completely independent from each other - they include spontaneous paralellism. It is evident that if the application needs maximal precision then the complicated QTP algorithm is not fast enough we suppose because of the round off errors.

The curves marked by A represent a mixed construction, we made a special form of preconditioning: we used in the upper and in the middle block tridiagonal preconditioning, and in the other blocks quadrant tridiagonal one. These latter contain other information too, than the normal QTP. So this special one, and the normal QTP are added to each other. This construction requires even more operations in one step, and of course converges much faster. This alogorithm can be interpreted as an alternating domain preconditioner.

If the differential operator is not so simple as a laplace, the iteration number increases - see on fig.5, and in these cases the quadrant tridiagonal preconditioner can help more.

The size of A determines the block structure. Namely: if we use n grid points in one dimension the order of A will be n^2 , and will contain $n * n$ blocks. As an evident way of paralellisation we can divide the $2k = n$ dimensional blokk matrix problem into k n-dimensional subproblems which can be solved in parallel.

Let's try to divide the structure of the matrix in an other way. On fig.5. we can see the result if instead of the original n=20 dimensional subproblems we use 4,10,25,30,40,80 dimensional bloks as subproblems. The convergence of this algorithm slows down. The only remarkable thing is visible on the second graph, where the various QT preconditioner block lengths can be seen. When the block length is only 4, the convergence speed is the half of the normal, but we can use 5 times more processors (if there are enough).

We must add that the efficience of these algorithms is good , these algorithms terminate faster, than the theoretical estimation. We have got approximately the same runtime values, independent from the preconditioning technique but we hope that the runtime values will show a significant dependence on the parallel computers with different internal structure.

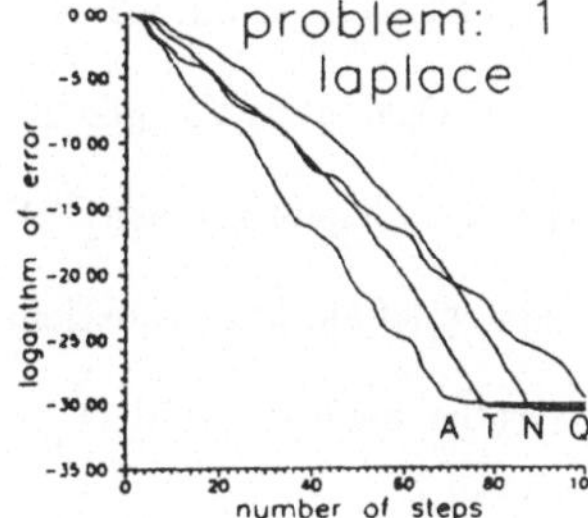

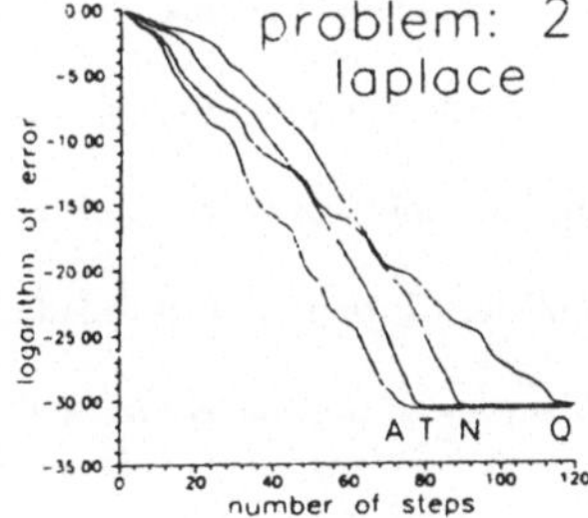

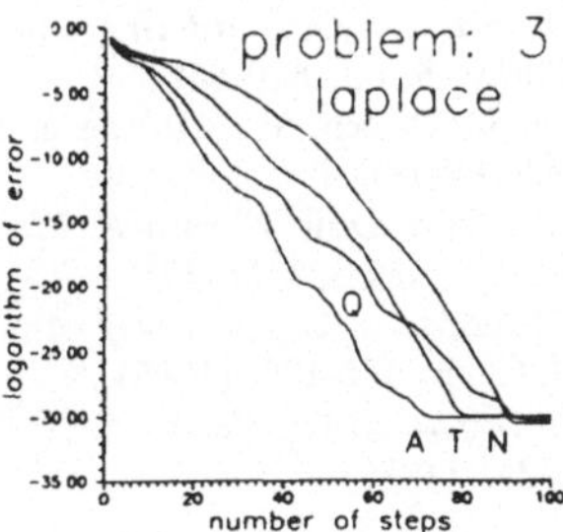

Fig. 4.

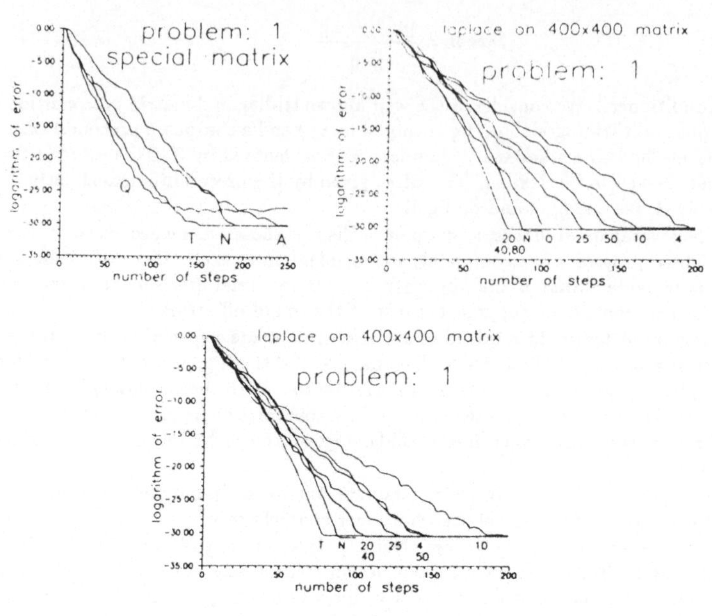

Fig. 5.

References

[1] Evans, D. J. , Hadjidimos, A. and Noutsos, D. , The parallel solution of banded linear equations by the new quadrant interlocking factorization (I.Q.F) method. International Journal of Computational Mathematics, 9. (1981).

[2] Evans, D. J., Parallel numerical algorithm for linear systems. In Parallel Processing Systems (ed D. Evans) Cambridge University Press, (1982).

[3] Evans, D. J., Parallel S.O.R. iterative methods. Parallel Computing, 1(1), (1984).

[4] O'Leary, D. , Ordering schemes for parallel processing of certain mesh problems. SIAM J. Sci. Stat. Comp. 5, pp. 620-632.

[5] Hackbusch, W., On the multigrid method applied to difference equations. Computing 20. pp. 291-306, (1978).

[6] Iain S. Duff, Gerard A. Meurant, The effect of ordering on preconditioned conjugate gradient. BIT 29. pp. 635-657, (1989).

[7] Axelsson, O., A survey of preconditioned iterative methods for linear system of algebraic equations. BIT, pp. 166-187, (1985).

[8] Axelsson, O., Vasilevski, P.S., A survey of multilevel preconditioned iterative methods. BIT 29, pp. 69-93, (1989).

A new approximate factorization method [1]

Yvan Notay [2]
Service de Métrologie Nucléaire
Université Libre de Bruxelles (C.P. 165)
50, Av. F.D. Roosevelt, B-1050 Brussels, Belgium.
email : ynotay@ulb.ac.be

Summary

A new incomplete factorization method is presented, differing from the previous ones by the way in which the diagonal entries of the triangular factors are defined. A comparison is given with other basic incomplete factorization methods, displaying the superiority of the new one, particularly for systems arising from anisotropic elliptic PDEs.

1. Introduction

The precondioned conjugate gradient method with preconditioning obtained by incomplete factorization of the system matrix is now a popular technique for the iterative solution of large sparse linear systems. For systems arising from the discretisation of second order elliptic PDEs, it is generally admitted that so-called "modified" factorizations are more efficient than the standard (or "unmodified") Incomplete Cholesky (IC)

[1]The present work was supported by the "Programme d'impulsion en Technologie de l'Information", financed by Belgian State, under contract No. IT/IF/14.

[2]Supported by the "Fonds National de la Recherche Scientifique", Chargé de recherches.

factorizations. In these methods, the diagonal part of the upper triangular factor is computed so that the preconditioner satisfies a (generalized) row-sum criterion. The version according to which this criterion is satisfied exactly sufffers however from an unreliable behaviour of its associate largest eigenvalue(s) (see e.g. [12, 16]), but this drawback may generally be avoided by apropriately perturbing the factorization algorithm so as to satisfy a prescribed upper eigenvalue bound.

A first such "perturbed" method is that initiated by Axelsson [1], generalized by Gustafsson [8], and whose algorithmic or *dynamic* version, first developped by Axelsson and Barker [3], has been reformulated in a more convenient fashion by Beauwens [6]. We shall refer to this method as DMIC (Dynamic Modified Incomplete Cholesky), the name MIC being reserved to the unperturbed version (row-sum criterion satisfied exactly).

Another interesting approximate factorization method is the Relaxed Incomplete Cholesky (RIC) method introduced by Axelsson and Lindskog [4]. It may also be seen as a perturbed modified incomplete factorization, the associate largest eigenvalue satisfying an upper bound similar to that of the DMIC method.

The method we shall present here is also such a perturbed approximate factorization method, based on a new upper eigenvalue analysis developped in [15]. It may be viewed as a dynamic versoin of the RIC method and, for that reason, we call it DRIC (Dynamic Relaxed Incomplete Cholesky).

As usual when dealing with approximate factorization methods, we assume that the system matrix is a Stieltjes matrix (i.e. positive definite with nonpositive offdiagonal entries), more general cases arising from the discrete PDE context being included through the reduction techniques presented in [3, 7, 11]. To clarify the presentation, we further restrict ourselves to diagonally dominant matrices, the general Stieltjes case being included with little handling, see [15]. Note also that, according to the results in [10], the case of consistent semidefinite systems is included with practically no change, see [15] for details.

2. Algorithms and basic properties

All above mentioned methods differing only by the way in which some parameters are chosen, we present them below in a single algorithm with different options. We use a pseudo programming language to clarify the presentation.

ALGORITHM

Initialise : $u_{ij} := a_{ij}$ *for all* $1 \leq i \leq j \leq n$, *and choose the paremeters* ω *and* τ *associated to respectively RIC and DMIC or DRIC such that*

$$\begin{cases} -1 \leq \omega < 1 & \text{for RIC} \\ 0 < \tau < 1 & \text{for DMIC} \\ 0 \leq \tau < 1 & \text{for DRIC} \end{cases} .$$

Execute then, for $k = 1, \ldots, n-1$:

$$\tau_0 := \frac{-\sum_{i>k} u_{ki}}{u_{kk}}$$

$$\text{if} \quad IC \quad : \quad \omega_c := 0$$
$$\text{if} \quad RIC \quad : \quad \omega_c := \omega$$
$$\text{if} \quad MIC \quad : \quad \omega_c := 1$$
$$\text{if} \quad DMIC \quad : \quad \{ \ \omega_c := 1 \ \text{if} \ \tau_0 > \tau \ : \quad u_{kk} := -\tau^{-1} \sum_{i>k} u_{ki}$$
$$\text{if} \quad DRIC \quad : \quad \begin{cases} \text{if} \ \tau_0 \leq \tau \ : & \omega_c := 1 \\ \text{otherwise} \ : & \omega_c := 2\tau/\tau_0 - 1 \end{cases}$$

for all $i > k$ *such that* $u_{ki} \neq 0$:

$$u_{ii} := u_{ii} - \frac{u_{ki}^2}{u_{kk}}$$

for all $j > i$ *such that* $u_{kj} \neq 0$:

if fill-in is allowed in position (i,j) :

$$u_{ij} := u_{ij} - \frac{u_{ki} u_{kj}}{u_{kk}}$$

otherwise :
$$\begin{cases} u_{ii} := u_{ii} - \omega_c \frac{u_{ki} u_{kj}}{u_{kk}} \\ u_{jj} := u_{jj} - \omega_c \frac{u_{ki} u_{kj}}{u_{kk}} \end{cases} .$$

It is seen that the IC method is the same as the RIC one with $\omega = 0$ while RIC with $\omega = -1$ and DRIC with $\tau = 0$ are identical.

$A = (a_{ij})$ being a (nonstrictly) diagonally dominant Stieltjes matrix, the algorithm cannot fail and it results an upper triangular matrix $U =$

(u_{ij}) whose regularity is ascertained except for MIC factorizations (see then [3, 10] for existence criteria). Letting

$$B = U^t P^{-1} U \tag{1}$$

where $P = diag(U)$ be the associate preconditioner, it further follows that

$$\nu_{\max}(B^{-1}A) \leq \begin{cases} \frac{2}{1-\omega} & \text{for RIC} \\ \frac{1}{1-\tau} & \text{for DMIC \& DRIC .} \end{cases}$$

The upper bound for DMIC is a classical resut; for (D)RIC, it follows from a algebraic result in [15].

None of the methods DMIC, RIC or DRIC avoids parameter dependency, but a practical rule for the determination of the parameter τ or ω is provided by :

$$\tau = 1 - \xi h_0 \tag{2}$$

for the DMIC & DRIC methods and

$$\omega = 1 - \delta h_0 \tag{3}$$

for the RIC method, where ξ and δ are near 1 and

$$h_0 = \frac{hS}{4V} \tag{4}$$

where h denotes the average mesh size, V the area/volume of the domain Ω in which the PDE is solved and S the lenght/area of its boundary (note that $h_0 = h$ if Ω is the unit square). It follows from numerical experiments that these rules are nearly optimal, see [4, 13, 15].

All three methods satisfy thus a similar upper eigenvalue bound. Their relative performance depends hence essentially on the behaviour of the lowest eigenvalues. Assuming a natural ordering of the unknows, it is well known that the lowest eigenvalues associate to the DMIC preconditioner are $\mathcal{O}(1)$, see e.g. [3, 5, 9]. In [13], we further prove that the number of iterations to reduce the relative error by a factor ε is effectively bounded by $c h_0^{-\frac{1}{2}} \ln \varepsilon^{-1}$ with c not too far from unity provided that the pth (nonzero) eigenvalue of the *unpreconditioned* system satifies

$$h_0^{-2} \lambda_{\min}^{(p)}(D^{-1}A) \geq c_p \tag{5}$$

(where $D = diag(A)$) for some c_p near from 1 with p small integer [3]. As far as we know, this condition is fulfilled by isotropic problems. But,

---

[3]this result is based on a bound on the convergence rate of the conjugate gradients which is shown in [14] to be reliable in presence of rounding errors.

see [15], it is generally not satified by (strongly) anisotropic problems, except in "model" circumstances, with constant coefficients and Dirichlet boundary conditions.

The behaviour of RIC and DRIC is discussed in [15]. For isotropic problems, the RIC method is less interesting than the DMIC one for small h_0, due to a less favourable asymptotic behaviour. But, for strongly anisotropic problems and realistic h_0, it has on the contrary a better and even far better behaviour.

In this context, the first attractive property of the DRIC method is that it behaves essentially like DMIC for isotropic problems, and like RIC for (purely) anisotropic problems. The DRIC method presents thus the usefull feature of automatically shifting to the most interesting method, DMIC or RIC, depending on the case at hand. This is of particular interest for moderately anisotropic problems, for which it would be difficult to predict which from the DMIC or the RIC method should behave better. Further, it is show that an *effective* improvement over both previous methods is obtained for problems presenting isotropic and anisotropic regions as well.

3. Numerical results

We have tested the different methods on the linear systems resulting from the 5-point finite difference approximation with uniform mesh size h of :

$$- \partial_x a_x \partial_x \, u \; - \; \partial_y a_y \partial_y \, u \; = \; f$$

on the unit square, with :

$$\begin{cases} u = 0 & \text{for } 0 \le x \le 1,\, y = 0 \\ \partial_n u = 0 & \text{on the remaining part of the boundary} \end{cases}$$

$$\begin{cases} a_x = 1 \\ a_y = \begin{cases} \zeta & \text{in } (\tfrac{1}{4}, \tfrac{3}{4}) \times (\tfrac{1}{4}, \tfrac{3}{4}) \\ 1 & \text{elsewhere} \end{cases} \end{cases}$$

$$f = \begin{cases} 100 & \text{in } (\tfrac{1}{4}, \tfrac{3}{4}) \times (\tfrac{1}{4}, \tfrac{3}{4}) \\ 0 & \text{elsewhere} \end{cases}$$

where ζ is a parameter.

We consider the generalized SSOR (no fill-in allowed) preconditioners associate with the lexicographic ordering (starting at the bottom left corner). The parameters τ (DMIC & DRIC) and ω (RIC) are chosen according to respectively (2) and (3) with $\xi = 1$ for DMIC, $\delta = 1$ for RIC and $\xi = 2$ for DRIC, these choices being the most appropriate following the experimental results in [4, 13, 15]. For completeness, we include the results obtained with Gustafsson method [8] which we call Statically Perturbed Modified Incomplete Cholesky (SPMIC) because it is similar to DMIC except that it uses predetermined or "static" pertubations; according to Gustafsson suggestion [4], we use $\xi_1 = 4$, $\xi_2 = 1$ and for $\mathcal{N}_2$ the set of gridnodes on the lines $x = 0$, $0 \leq y \leq 1$ and $y = \frac{1}{4}$, $\frac{1}{4} \leq x \leq \frac{3}{4}$.

In Fig. 1-2, we have plotted the number of iterations to reduce the relative residual error by a factor 10^{-6} as a function of ζ when using the PCG algorithm with zero initial approximation. It is seen that all three methods are nearly equivalent for the "model" problem which corrresponds to $\zeta = 1$. For $\zeta > 1$, DMIC behaves better than RIC until its robustness condition (5) is satisfied but is very bad for higher values of ζ, while DRIC brings in any case a noteworthy improvement over the best of both other methods.

4. Multigrid smoothing

We have not tested the method as multigrid smoother, but we have observed that the upper bound

$$\nu_{\max}(B^{-1}A) \leq \frac{1}{1 - \tau}$$

is very sharp when one uses DRIC with small τ (say $\tau = .5$ or $\tau = 0$). Further, many eigenvalues are clustered around this value, and it is expected that these modes correspond to "local" modes. It follows that $C = (1 + \alpha)(1 - \tau)B^{-1}$ with α small nonnegative number should be an efficient multigrid smoother, each iteration of the type

$$x_{k+1} = x_k - (1 - \tau)(1 + \alpha)B^{-1}(b - Ax_k)$$

---

[4]private communication

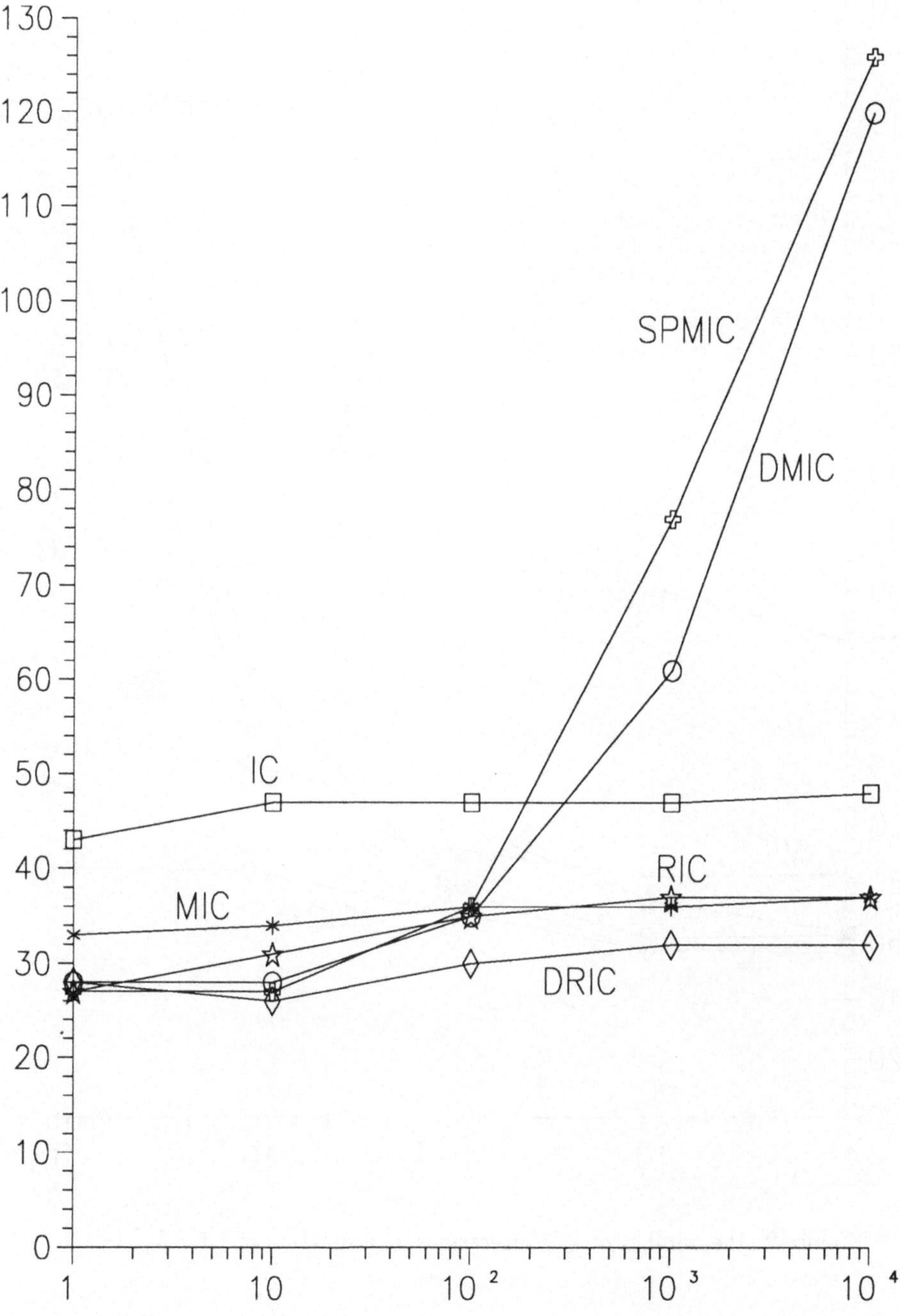

Figure 1: the number of PCG iterations as a function of ζ for $h_0^{-1} = 32$.

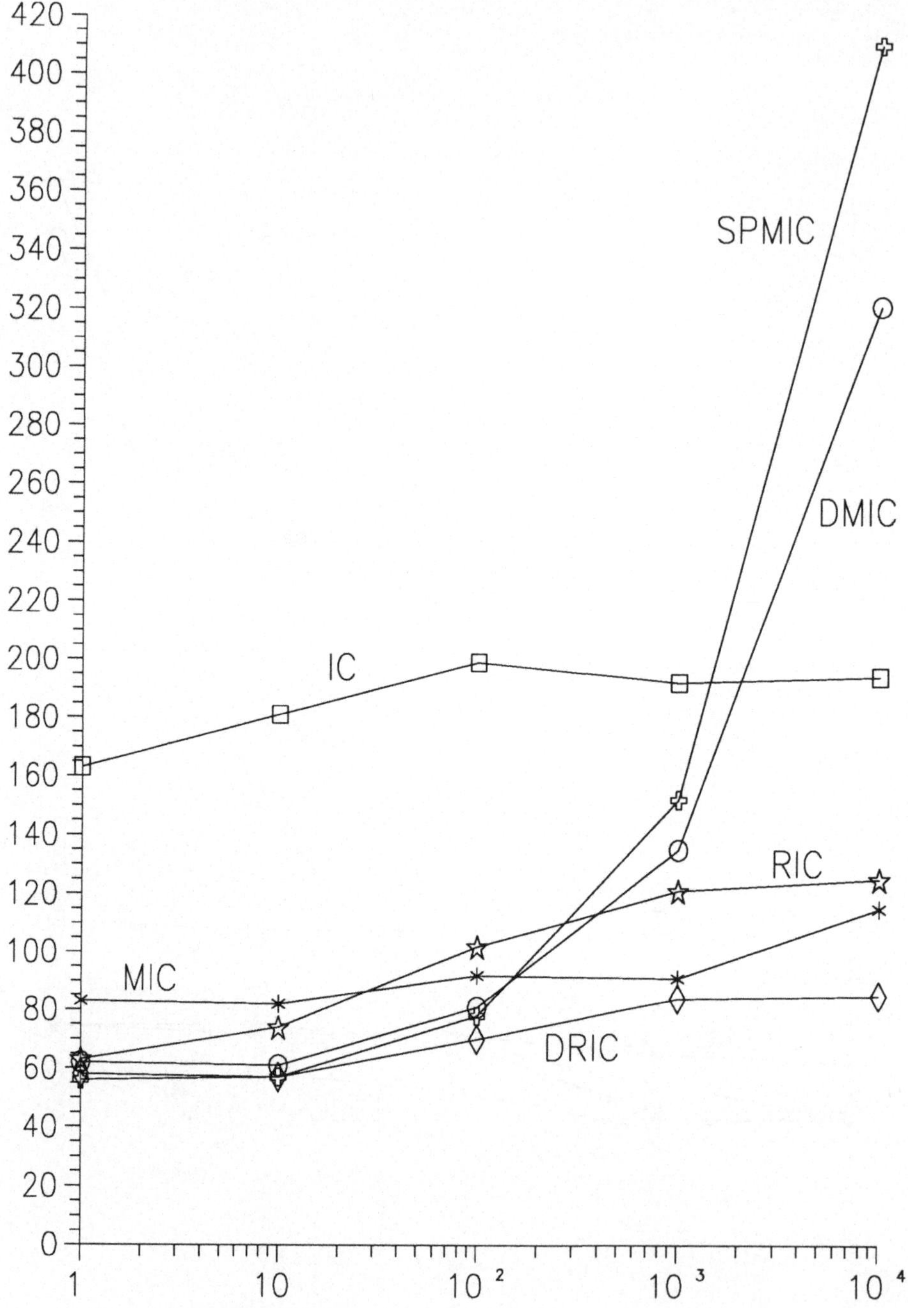

Figure 2: the number of PCG iterations as a function of ζ for $h_0^{-1} = 128$.

reducing the error corresponding to all modes in $[\frac{\beta}{1-\tau}, \nu_{max}]$ by a factor at most $\lambda = \max\left(\alpha, \frac{1-\beta}{1+\alpha}\right)$.

Theoretical and experimental results on RIC smoothing are reported in [17] for a model anisotropic problem. In strongly anisotropic cases, DRIC reduces to RIC with $\omega = 2\tau - 1$, so that these results concern the method suggested above, except that the correction factor $(1-\tau)(1+\alpha)$ is omitted. It is interesting to note in this respect that the optimal smoothing properties are obtained with $\omega \approx -\frac{1}{2}$ which corresponds to $1-\tau \approx \frac{3}{4}$, while $1+\alpha = \frac{4}{3}$ seems to be reasonable. The above considerations suggest then to examine whether a correction factor of about $\frac{4}{3}(1-\tau)$ help to obtain results less dependent on the parameter choice, and further whether the optimal parameter is less dependent on the anisotropy ratio by using DRIC rather than RIC.

References

[1] O. Axelsson, *A generalized SSOR method*, BIT, 13 (1972), pp. 443–467.

[2] O. Axelsson, *On iterative solution of elliptic difference equations on a mesh connected array of processors*, J. High Speed Comput., 1 (1989), pp. 165–184.

[3] O. Axelsson and V. Barker, *Finite Element Solution of Boundary Value Problems. Theory and Computation*, Academic Press, New York, 1984.

[4] O. Axelsson and G. Lindskog, *On the eigenvalue distribution of a class of preconditioning methods*, Numer. Math., 48 (1986), pp. 479–498.

[5] R. Beauwens, *Lower eigenvalue bounds for pencils of matrices*, Lin. Alg. Appl., 85 (1987), pp. 101–119.

[6] R. Beauwens, *Modified incomplete factorization strategies*, in Preconditioned Conjugate Gradient Methods, O. Axelsson and L. Kolotilina, eds., Lectures Notes in Mathematics No. 1457, Springer-Verlag, 1990, pp. 1–16.

[7] R. BEAUWENS AND R. WILMET, *Conditioning analysis of positive definite matrices by approximate factorizations*, J. Comput. Appl. Math., 26 (1989), pp. 257–269.

[8] I. GUSTAFSSON, *Modified Incomplete Cholesky (MIC) methods*, in Preconditioning Methods. Theory and Applications, D. Evans, ed., Gordon and Breach, New York-London-Paris, 1983, pp. 265–293.

[9] Y. NOTAY, *Incomplete factorization of singular linear systems*, BIT, 29 (1989), pp. 682–702.

[10] Y. NOTAY, *Solving positive (semi)definite linear systems by preconditioned iterative methods*, in Preconditioned Conjugate Gradient Methods, O. Axelsson and L. Kolotilina, eds., Lectures Notes in Mathematics No. 1457, Springer-Verlag, 1990, pp. 105–125.

[11] Y. NOTAY, *Résolution iterative de systèmes linéaires par factorisations approchées*, PhD thesis, Service de Métrologie Nucléaire, Université Libre de Bruxelles, Brussels, Belgium, 1991.

[12] Y. NOTAY, *Upper eigenvalue bounds and related modified incomplete factorization strategies*, in Iterative Methods in Linear Algebra, R. Beauwens and P. de Groen, eds., North-Holland, 1992, pp. 551–562.

[13] Y. NOTAY, *On the robustness of modified incomplete factorization methods*, Inter. J. Comp. Math., to appear, (1991).

[14] Y. NOTAY, *On the convrgence rate of the conjugate gardients in presence of rounding errors*, Numer. Math., submitted, (1991).

[15] Y. NOTAY, *A dynamic version of the RIC method*, submitted for publication.

[16] H. VAN DER VORST, *The convergence behaviour of preconditioned CG and CG-S*, in Preconditioned Conjugate Gradient Methods, O. Axelsson and L. Kolotilina, eds., Lectures Notes in Mathematics No. 1457, Springer-Verlag, 1990, pp. 126–136.

[17] G. WITTUM, *On the robustness of ILU-smoothing*, SIAM J. Sci. Statist. Comput., 10 (1989), pp. 699–717.

The use of sparse matrix techniques for solving the incompressible Navier–Stokes equations

A. van der Ploeg & F.W. Wubs,
Department of Mathematics , University of Groningen,
P.O. Box 800, 9700 AV Groningen, the Netherlands

Abstract

The system of non linear equations arising in the calculation of fluid flows governed by the incompressible Navier–Stokes equations is considered. When using Newton or Picard iteration a system of linear equations of the form $Ax = b$ has to be solved several times. Some sparse matrix techniques are introduced to solve these matrix equations.

Various computations of the laminar flow over a backward facing step are made. We do not take advantage of the special geometry of this problem and no restrictions are made with respect to the sparsity pattern of the coefficient matrices. The methods are tested for two different discretizations of the convective terms (upwind discretization and central differences) and for Reynoldsnumbers equal to 150 and 500.

1 Introduction

In this paper the linear systems arising in the calculation of fluid flows governed by the incompressible Navier–Stokes equations are considered. The time independent flow in two dimensions is governed by the equations

$$u\frac{\partial u}{\partial x} + v\frac{\partial u}{\partial y} = -\frac{\partial p}{\partial x} + \nu(\frac{\partial^2 u}{\partial x^2} + \frac{\partial^2 u}{\partial y^2}), \tag{1}$$

$$u\frac{\partial v}{\partial x} + v\frac{\partial v}{\partial y} = -\frac{\partial p}{\partial y} + \nu(\frac{\partial^2 v}{\partial x^2} + \frac{\partial^2 v}{\partial y^2}), \tag{2}$$

$$\frac{\partial u}{\partial x} + \frac{\partial v}{\partial y} = 0, \tag{3}$$

where u and v are the velocities in the $x-$ and $y-$direction respectively, p is the pressure and ν the kinematic viscosity of the fluid. These equations are discretised on staggered grids with u, v and p defined on different locations as shown in Fig. I.

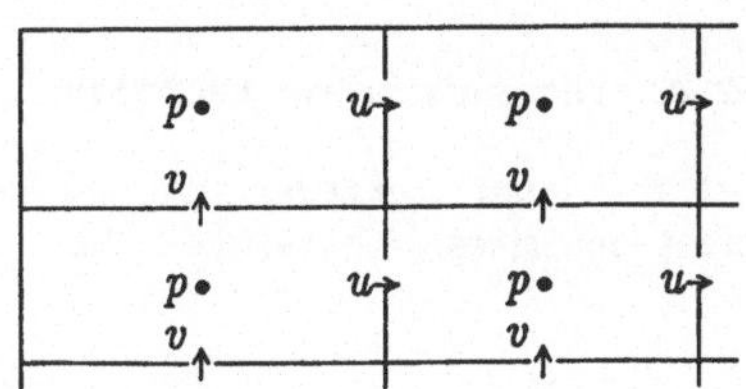

Fig. I The Marker and Cell method.

113

To deal with the non linearity of (1)–(3) we have to use an iterative method. In order to obtain systems of linear equations which are relatively easy to solve we first consider the Picard iteration. Suppose u^n, v^n and p^n are the results after n iterations. Then u^{n+1}, v^{n+1} and p^{n+1} are obtained from the equations

$$u^n\frac{\partial u^{n+1}}{\partial x} + v^n\frac{\partial u^{n+1}}{\partial y} = -\frac{\partial p^{n+1}}{\partial x} + \nu(\frac{\partial^2 u^{n+1}}{\partial x^2} + \frac{\partial^2 u^{n+1}}{\partial y^2}), \tag{4}$$

$$u^n\frac{\partial v^{n+1}}{\partial x} + v^n\frac{\partial v^{n+1}}{\partial y} = -\frac{\partial p^{n+1}}{\partial y} + \nu(\frac{\partial^2 v^{n+1}}{\partial x^2} + \frac{\partial^2 v^{n+1}}{\partial y^2}), \tag{5}$$

$$\frac{\partial u^{n+1}}{\partial x} + \frac{\partial v^{n+1}}{\partial y} = 0. \tag{6}$$

At every step of the iterative method defined by these equations a linear system $Ax = b$ of the form

$$\begin{pmatrix} C_1 & 0 & G_x \\ 0 & C_2 & G_y \\ D_x & D_y & 0 \end{pmatrix} \begin{pmatrix} u^{n+1} \\ v^{n+1} \\ p^{n+1} \end{pmatrix} = \begin{pmatrix} b_1^{n+1} \\ b_2^{n+1} \\ b_3^{n+1} \end{pmatrix} \tag{7}$$

has to be solved, where G_x and G_y are approximately equal to $-D_x$ and $-D_y$ respectively. In the sequel of this paper we will write (7) as

$$\begin{pmatrix} A_{11} & G \\ D & 0 \end{pmatrix} \begin{pmatrix} u^{n+1} \\ p^{n+1} \end{pmatrix} = \begin{pmatrix} b_1^{n+1} \\ b_2^{n+1} \end{pmatrix}. \tag{8}$$

When using an upwind finite difference scheme for the convective terms and a standard discretization for the Laplace equation the block A_{11} is an M–matrix so the construction of an incomplete decomposition of it is straightforward (see e.g. [1]). To solve this system of linear equations one can use an iterative method suitable for non–symmetric systems like CGSTAB [2] or GMRES(M) [3]. In this context such a method to solve the linear systems can be looked at as an inner iteration within the Picard iteration. Using an iterative method has some advantages over direct solution of the linear systems. First of all iterative methods can take advantage of good starting values for u^{n+1} and p^{n+1}. Secondly, the stop criterion of the inner iteration process at stage n of the outer iteration process can be adapted according to the accuracy of u^{n+1} and p^{n+1} that can be obtained at stage n of the outer iteration.
In order to improve the convergence behavior of the inner iteration we apply a preconditioning technique which is described in the next section.

2 A preconditioning technique

The speed of convergence of the inner iteration strongly depends on the eigenvalue distribution of the coefficient matrices. Therefore, the inner iteration is applied on the equivalent system

$$C^{-1}\begin{pmatrix} I & 0 \\ 0 & C_p \end{pmatrix}\begin{pmatrix} A_{11} & G \\ D & 0 \end{pmatrix}\begin{pmatrix} u^{n+1} \\ p^{n+1} \end{pmatrix} = C^{-1}\begin{pmatrix} I & 0 \\ 0 & C_p \end{pmatrix}\begin{pmatrix} b_1 \\ b_2 \end{pmatrix}.$$

One can think of the matrix

$$\begin{pmatrix} I & 0 \\ 0 & C_p \end{pmatrix}$$

to be a sort of pre–preconditioner. The function of it is to obtain a coefficient matrix for which one can easily construct a good preconditioner C. We try to choose the block C_p such that $C_p D \approx D A_{11}$, so C should be an approximation of the matrix

$$\begin{pmatrix} A_{11} & G \\ C_p D & 0 \end{pmatrix} \approx \begin{pmatrix} A_{11} & G \\ D A_{11} & 0 \end{pmatrix}.$$

Since the block A_{11} is an M–matrix we can easily construct an incomplete decomposition $L_1 U_1$ of it which gives the following possible choice for C

$$C = \begin{pmatrix} I & 0 \\ D & L_2 U_2 \end{pmatrix}\begin{pmatrix} L_1 U_1 & G \\ 0 & I \end{pmatrix} \approx \begin{pmatrix} L_1 U_1 & G \\ D L_1 U_1 & 0 \end{pmatrix}, \tag{9}$$

where $L_2 U_2$ is is an incomplete decomposition of $-DG$. The construction of L_2 and U_2 is relatively easy because the matrix product DG is approximately equal to the coefficient matrix which arises after a standard discretization of the Laplace equation.

The choice of the matrix C_p.
As mentioned before, C_p should be chosen such that

$$C_p D \approx D A_{11}.$$

The block A_{11} has the form

$$\begin{pmatrix} C_1 & 0 \\ 0 & C_2 \end{pmatrix},$$

where C_1 and C_2 are discretizations at u– and v–points of the same differential operator

$$u\frac{\partial}{\partial x} + v\frac{\partial}{\partial y} - \nu\left(\frac{\partial^2}{\partial x^2} + \frac{\partial^2}{\partial y^2}\right),$$

so C_1 and C_2 are approximately the same. Since D comes from the discretization of first order derivatives in the x– and y–direction and the differential operators approximately commute when u and v are not strongly varying, the choice of C_p equal to C_1 or C_2 gives the desired property. In this paper C_p was taken equal to C_1. This choice has the nice property that the pre–preconditioner will get better when the mesh is refined, because in that case the matrix blocks will be more accurate discretizations of the differential operators and the choice of C_p is based on the fact that the latter approximately commute.

We do not make use of the geometry of the problem and no restriction is made with respect to the sparsity pattern of the blocks D, C_1 and C_2. As a consequence all the elements in the coefficient matrix and in the preconditioner have to be addressed indirectly.

3 Numerical experiments

To test the preconditioning techniques described above we compute the stationary flow in a two dimensional channel with a backward facing step. This so called Backward Facing Step problem has become a well known model problem to test the accuracy and efficiency of incompressible Navier–Stokes solvers [4]. The domain under consideration is shown in the next figure

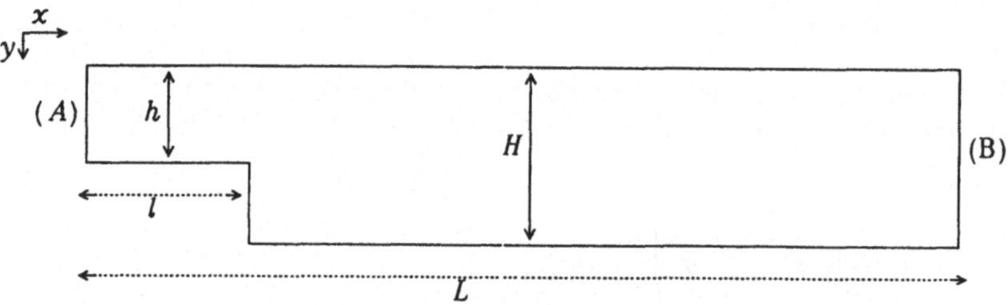

Fig. II The geometry of the Backward Facing Step problem.

where $h = 2$, $H = 3$, $L = 44$ and $l = 6$. At (A) a parabolic velocity profile is prescribed:

$$u(y) = \frac{4}{h^2}(h - y)y.$$

At the end of the channel the following boundary conditions are used:

$$\frac{\partial u}{\partial x} = 0 \text{ and } v = 0.$$

These boundary conditions are based on the hypothesis that (B) is far enough from the step for the flow to become parallel and not to change anymore along the x-direction. At all the other boundaries the impermeability and non–slip conditions are used: $u = 0$ and $v = 0$.

The next tables show the results of the preconditioning technique described earlier. Equations (4)–(6) are discretised on an equidistant grid using an upwind finite difference scheme for the convective terms. The grid is refined several times by reducing the mesh sizes in both the x- and y-direction. The first column shows the number of degrees of freedom, the third column the total number of inner iterations and the fourth column the number of Picard iterations. Of course this number increases with the Reynoldsnumber. The Picard iteration is stopped when the maximum norm of the difference between two succeeding iterations is less than 10^{-4}. The stop criteria for CGSTAB and GMRES are also based on the maximum norm of this difference. The second column shows the number of entries in the preconditioner. The last column gives the CPU–time in seconds measured on a HP 720 work station.

Table 1 Upwind finite differences combined with
CGSTAB and Picard iteration. Re = 150.

degrees of freedom	Number of entries	Mat. Vec. multipl.'s	Outer iterations	CPU-time
1512	21017	110	12	2.1
3402	53988	178	13	8.2
6048	101202	222	15	19.2
9450	164191	294	17	40.2

Table 2 Upwind finite differences combined with
GMRES(20) and Picard iteration. Re = 150.

degrees of freedom	Number of entries	Mat. Vec. multipl.'s	Outer iterations	CPU-time
1512	21090	82	12	2.0
3402	54132	130	15	8.4
6048	102175	147	15	18.9
9450	164671	139	15	31.7

Since the number of GMRES iterations required per outer iteration step is less than 20, it is not necessary to restart the inner iteration. The number of non–zero entries in the preconditioning is approximately 17 times the number of degrees of freedom. This is very low compared with the number of non–zero entries which are needed when using a complete LU–decomposition.

GMRES appears to be approximately as expensive as CGSTAB on a scalar machine like the HP 720 work station. The fact that the convergence behavior of CGSTAB can be irregular makes it difficult to obtain an optimal stopping criterion for the inner iteration. Secondly, CGSTAB has the disadvantage that two matrix vector multiplications are needed per iteration but when using GMRES much work has to be done for obtaining an orthonormal basis for the Krylov–subspace. However, these extra computations consist mainly of calculating inner products which can be vectorized very well. Therefore, from now on we only use GMRES(M) as an inner iteration. A drawback of GMRES(M) is that computer storage demands become higher with increasing M. When the value of M is chosen too small the convergence behavior is very bad or GMRES(M) may not converge at all.

Using central differences for the convective terms.
We will now consider the more realistic case in which a more accurate discretization is used for the convective terms. Consider the following three grid points with corresponding function values ϕ_-, ϕ_0 and ϕ_+:

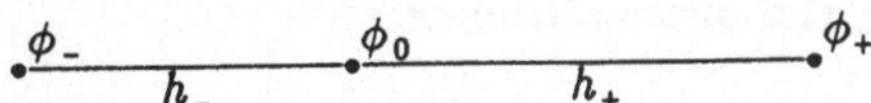

In the second grid point we want to have a second order accurate discretization of $\partial\phi/\partial x$. One can use the first term in the right–hand side of the following formula

$$\frac{\partial \phi}{\partial x} = \frac{\phi_+ - \phi_-}{h_- + h_+} - \tfrac{1}{2}(h_+ - h_-)\frac{\partial^2 \phi}{\partial x^2} - \frac{1}{6}\frac{h_+^3 + h_-^3}{h_+ + h_-}\frac{\partial^3 \phi}{\partial x^3} + \cdots .$$

On first sight the error term with $(h_+ - h_-)$ looks a first order term, but when the grid is obtained from a transformation $x_i = f(\xi_i)$, where the points ξ_i form an equidistant grid with mesh–size Δ, we have

$$h_+ = x_+ - x_0 = \Delta f'(\xi_0) + \tfrac{1}{2}\Delta^2 f''(\xi_0) + \cdots,$$

$$h_- = x_0 - x_- = \Delta f'(\xi_0) - \tfrac{1}{2}\Delta^2 f''(\xi_0) + \cdots$$

and one can readily see that the difference $(h_+ - h_-)$ is in fact a second order term. The term $u\partial u/\partial x$ will therefore be discretised as

$$u_0 \frac{u_+ - u_-}{h_- + h_+} .$$

The other convective terms are discretised similarly. This leads to a system of linear equations in which all diagonal entries are positive. To make sure that the mesh Peclet numbers are not large where the flow is strongly varying, we use non–uniform rectangular grids. The results for Reynoldsnumber equal to 500 are shown in table 3 (when the number of degrees of freedom is 1056 GMRES(20) does not converge).

Table 3 Central differences combined with GMRES(20)
and Picard iteration. Re = 500. Non–uniform rectangular grid.

degrees of freedom	Number of entries	Mat. Vec. multipl.'s	Outer iterations	CPU–time
1056				
2376	27781	3502	85	142.4
4224	53801	3265	92	265.4
6600	89316	3819	106	560.8

The construction of an incomplete factorization of A_{11} does not fail even when this block is not an M–matrix. The number of outer iterations can easily be reduced by using Newton iteration as soon as one has obtained an approximate solution which is close enough to the exact solution. However, at every step of the outer iteration GMRES(20) needs more than 35 matrix–vector multiplications to solve the linear system. In order to reduce this number one has to use an improved preconditioning technique.

4 Improving the preconditioner

Again we use the pre–preconditioner which is introduced earlier. In this section we try to develop a preconditioner which is a more accurate approximation of the coefficient matrix

$$\begin{pmatrix} A_{11} & G \\ C_p D & 0 \end{pmatrix}$$

than the matrix product in (9). Assuming that L_1U_1 is a proper incomplete decomposition of A_{11} and L_2U_2 is an incomplete factorization of $-C_pD(L_1U_1)^{-1}G$, the matrix product

$$\begin{pmatrix} L_1U_1 & 0 \\ C_pD & L_2U_2' \end{pmatrix} \begin{pmatrix} I & (L_1U_1)^{-1}G \\ 0 & I \end{pmatrix} \approx \begin{pmatrix} L_1U_1 & G \\ C_pD & 0 \end{pmatrix}$$

is a good preconditioner for the coefficient matrix. In practice it is very difficult to construct a preconditioner for the block $-C_pD(L_1U_1)^{-1}G$ because $(L_1U_1)^{-1}$ is not available. Instead a block $\hat{D}$ is constructed which is approximately equal to $C_pD(L_1U_1)^{-1}$. Since C_p is chosen such that C_pD is approximately equal to DA_{11}, $\hat{D}$ will be similar to the block D. The construction of an incomplete decomposition of $-\hat{D}G$ is expected to be relatively cheap because this block is very similar to the coefficient matrix arising from a standard discretization of the Laplace equation. Solving $\hat{D}$ from $C_pD = \hat{D}L_1U_1$ can be done in two steps. First the block $\tilde{D}$ is solved from

$$C_pD = \tilde{D}U_1 \tag{10}$$

and then $\hat{D}$ is solved from

$$\tilde{D} = \hat{D}L_1. \tag{11}$$

The matrix $\tilde{D}$ can be constructed row by row. Suppose that the elements $\tilde{d}_{ij}$ have been constructed for $j < k$. The element $\tilde{d}_{ik}$ can be constructed from (10):

$$rl_{ik} = \sum_{j=1}^{k} \tilde{d}_{ij}u_{jk},$$

in which rl_{ik} represents the (i,k)–element of the matrix product C_pD and u_{ik} represents the corresponding entry of U_1. Assuming that all diagonal entries of U_1 are 1 the last equation can be written as

$$\tilde{d}_{ik} = rl_{ik} - \sum_{j=1}^{k-1} \tilde{d}_{ij}u_{jk}. \tag{12}$$

To be able to work with very large coefficient matrices the block $\tilde{D}$ should be sparse. This can be realized by solving (12) only approximately. Suppose max is the maximum absolute value of the entries in the i–th row of C_pD. When the absolute value of the right–hand side of (12) is less than ε times max the entry $\tilde{d}_{ik}$ is assumed to be zero. Herein ε is a threshold parameter which has to be chosen in advance. Solving $\hat{D}$ from (11) can be done similarly. As soon as row i of $\hat{D}$ has been constructed row i of L_2 and U_2 can be calculated. The construction of an incomplete decomposition of $-\hat{D}G$ as described above does not require the matrices $\hat{D}$ and $\tilde{D}$ to be stored in memory, because the construction of row i of L_2 and U_2 only requires the corresponding row of $\hat{D}$ and $\tilde{D}$.

Tables 4 and 5 give the results of this technique. The preconditioner was used in the following form:

$$C = \begin{pmatrix} I & 0 \\ \hat{D} & L_2 U_2 \end{pmatrix} \begin{pmatrix} L_1 U_1 & G \\ 0 & I \end{pmatrix}.$$

This form has the advantage that the matrix $(L_1 U_1)^{-1}$ has to be used only once when solving the system $C^{-1} x = b$. A disadvantage of this form of the preconditioner is that $\hat{D}$ has to be stored in memory. As $\hat{D}$ is approximately equal to D, this is not so bad. The second column gives the total number of entries necessary for the preconditioner.

Table 4 Central differences combined with GMRES(20)
and Picard iteration. Re = 150. Non-uniform rectangular grid.

degrees of freedom	Number of entries	Mat. Vec. multipl.'s	Outer iterations	CPU-time
1056	31934	109	19	4.8
2376	76864	150	20	15.3
4224	143888	167	21	32.2
6600	231455	184	21	58.0

Table 5 Central differences combined with GMRES(20)
and Picard iteration. Re = 500. Non-uniform rectangular grid.

degrees of freedom	Number of entries	Mat. Vec. multipl.'s	Outer iterations	CPU-time
1056	36421	700	63	22.2
2376	88950	602	90	59.4
4224	168888	643	83	126.8
6600	273875	665	74	216.5

From the results one can see that this preconditioning technique strongly reduces the number of matrix–vector multiplications required when applying one step of the Picard iteration although the number of entries in the preconditioner has increased with a factor three.

5 Conclusions

In this paper a preconditioner for the full system of equations arising after a linearisation of the Navier–Stokes equation is developed. A straightforward sparse incomplete factorization for this system is not possible. Therefore, we use a pre–preconditioner. After applying this pre–preconditioner to the system sparse incomplete factorization of the resulting linear system can be made, which results in

savings on CPU–time and storage. The preconditioning performs well for both upwind
and central finite differences.
We used both CGSTAB and GMRES. On scalar machines both are equally efficient,
although GMRES has the drawback of requiring more storage. However, on vector
computers, GMRES is more attractive since a large part of the computations in GMRES
consists of calculating inner products which vectorize quite well. The techniques
described in this paper have no restriction with respect to the sparsity pattern.
Hence, far more complicated geometries can be handled. The other side of this
generality is that indirect addressing is unavoidable, which is disadvantageous on
some vector computers. However, for regular geometries we can do without indirect
addressing and good performance on all types of super computers is expected.
Instead of Picard iteration also Newton's method can be used as an outer iteration.
In that case the results are similar.

Acknowledgments
This research was supported by Rijkswaterstaat, Rijswijk, the Netherlands. The
authors wish to thank Dr. E.F.F. Botta for many useful suggestions and stimulating
discussions.

References

1. J.A. Meijerink and H.A. van der Vorst, *'An iterative solution method for linear
   systems of which the coefficient matrix is a symmetric $M-matrix$'*, Math. Comput.,
   31, 148–162 (1977).
2. H.A. van der Vorst and P. Sonneveld, *'CGSTAB: a more smoothly converging variant
   of $CG-S$'*, Reports of the Faculty of Technical Mathematics and Informatics no.
   90–50, Delft (1990).
3. Y. Saad and M.H. Schultz, *'GMRES: a generalized minimal residual algorithm for
   solving non-symmetric linear systems'*, SIAM J. Sci. Stat. Comput., **7**, 856–869
   (1986).
4. K. Morgan, J. Periaux and F. Thomasset, *'Analysis of Laminar Flow over a Backward
   Facing Step'*, Notes on numerical fluid mechanics, Vol. **9** (1984).

THE CONSTRUCTION OF THE INTERPOLATION OPERATOR WITH I L U
DECOMPOSITION FOR ALGEBRAIC POSITIVE DEFINITE SYSTEMS

Constantin Popa

University of Constanta, Department of Mathematics

Bd.Mamaia,nr.124,Constanta 8700,Romania

Summary

We present in this paper two results concerning the convergence of the two-grid algebraic algorithm for arbitrary symmetric systems of linear equations which are also positive definite. We obtain these results using a special construction of the interpolation operator based on Gaussian elimination on a sub-matrix of the original system matrix. At the end of the paper we make also some remarks concerning the symmetric indefinite systems.

1. Introduction

We consider the two-grid algebraic method

$$\bar{x} = x + I_p^n A_p^{-1} I_n^p (b - Ax) \tag{1}$$

$$\bar{\bar{x}} = S\bar{x} + Tb \tag{2}$$

where A is a $n \times n$ symmetric matrix (and positive definite), the operator $I_p^n : \mathbb{R}^p \longrightarrow \mathbb{R}^n$ is the interpolation, S and T are relaxation operators, I_n^p , A_p are given by the variational conditions

$$I_n^p = (I_p^n)^t \tag{3}$$

$$A_p = I_n^p A I_p^n \tag{4}$$

and x is a given approximation of the exact solution of the system

$$Ax = b . \tag{5}$$

Let $\| \cdot \|_0$, $\| \cdot \|_1$,resp. $\| \cdot \|_2$ be norms on $\mathbb{R}^n$ given by the scalar products

$$\langle x , y \rangle_0 = \langle Dx , y \rangle , \langle x , y \rangle_1 = \langle Ax , y \rangle$$
$$\langle x , y \rangle_2 = \langle D^{-1}Ax,Ay \rangle \tag{6}$$

where $D = \text{diag}(A)$ and $\langle \, , \, \rangle$ is the Euclidian scalar product and $\|\cdot\|$ the associated Euclidian norm. We know the classical approximation property (AP) and smoothing property (SP) of the algebraic multigrid ([10])

$$\text{(AP)}: \quad (\exists)\,\beta > 0, \; \beta \, \| \bar{e} \|_1^2 \; \leq \; \| \bar{e} \|_2^2 \tag{7}$$

$$\text{(SP)}: \quad (\exists)\,\alpha > 0, \; \| \bar{\bar{e}} \|_1^2 \leq \| \bar{e} \|_1^2 - \alpha \| \bar{e} \|_2^2 \tag{8}$$

and the fact that, if (7) and (8) holds, then the two-grid method (1) – (2) converges with a convergence factor bounded in $\| \cdot \|_1$ – norm by $\sqrt{1 - \alpha\beta}$ ([10] , [4]). The smoothing property (8) is satisfied by the classical ω – Jacobi, Gauss-Seidel and S O R relaxations schemes (see [1] , [10] , [4] , [8]) for some classes of symmetric and positive definite matrices A (we have noted by $\bar{e}$, resp. $\bar{\bar{e}}$, the errors $\bar{x} - x^*$, resp. $\bar{\bar{x}} - x^*$). One of the most difficult problems of the algebraic multigrid is to obtain the approximation property (7). It is proved (see [10]) that the following condition

$$(\exists)\,b > 0., \; \min\left\{ \| u - I_p^n u_p \|_0^2 \; \Big| \; u_p \in R^p \right\} \leq b \cdot \| u \|_1^2 \tag{9}$$

implies (7) with β given by

$$\beta = 1 / b. \tag{10}$$

<u>Observation.</u> It can be proved ([4]) that the condition (9) must be verified only by vectors $u \in R^n$ which satisfy

$$I_n^p A u = 0 \tag{11}$$

condition which is valid, for example, if u is the error after the coarse – grid correction step. We shall prove in what follows conditions of the type (9) for a special construction of the interpolation operator I_p^n .

2. <u>The construction of I_p^n</u>

Let

$$A = \begin{bmatrix} A_1 & B \\ B^t & A_2 \end{bmatrix} \tag{12}$$

be a decomposition of A, A_1 and A_2 being symmetric matrices of dimension $n-p$, resp. $p, 1 \leq p < n$ (A_2 corresponds to the coarse grid – see also the last section of the present paper) We suppose that A_1 is strictly diagonaly dominant, thus

123

$$a_{ii} - \sum_{j \neq i} |a_{ij}| = \gamma_i > 0 \ , \ i = 1,\ldots,n-p \tag{13}$$

and let

$$0 < \mu_1 \leq \mu_2 \leq \ldots.. \leq \mu_{n-p} \tag{14}$$

be the eigenvalues of A_1 .From the Gershgorin theorem we know that

$$\mu_1 \geq \gamma = \min \{ \gamma_i \mid 1 \leq i \leq n-p \} . \tag{15}$$

In what follows we shal note by I_1 , I_2 ,I the identity on $\mathbb{R}^{n-p}$, $\mathbb{R}^p$, resp. $\mathbb{R}^n$ and by $u = [u^1 , u^2]^t$ the decomposition of a vector $u \in \mathbb{R}^n$ with respect to the decomposition $\mathbb{R}^n = \mathbb{R}^{n-p} \oplus \mathbb{R}^p$ .We shall also suppose that $D = \mathrm{diag}(A)$ satisfies

$$D = \alpha \cdot I \ , \alpha > 0 \tag{16}$$

We shall then define the interpolation operator I_p^n as follows

$$I_p^n = \begin{bmatrix} - A_1^{-1} B \\[2mm] I_2 \end{bmatrix} \tag{17}$$

$$I_p^n = \begin{bmatrix} -1/\alpha \cdot B \\[2mm] I_2 \end{bmatrix} . \tag{18}$$

<u>Remarks</u>. 1. In (17) the matrix A_1 must not be realy inverted.It can be proved that the product $A_1^{-1} B$ can be obtained by Gaussian elimination ([7])
2. The fomula (18) corresponds to the classical construction of the interpolation operator in the algebraic multigrid (see [1], [3], [10]).
3. Both constructions (17) and (18) are particular cases of an I L U decomposition applied to the matrix A_1 (in fact they are extreme cases – the best and the worst , see [5], [7]) .

3. The main results

<u>Theorem</u> 1. <u>If</u> $u \in \mathbb{R}^n$ <u>satisfies</u> (11) <u>and</u> I_p^n <u>is given by</u> (17) <u>then</u>

$$\langle A u , u \rangle = \langle A_1 u^1 , u^1 \rangle \geq \mu_1 \cdot \| u^1 \|^2 . \tag{19}$$

<u>Proof</u>. The relation (11) is equivalent with

$$\begin{bmatrix} -B^t A_1^{-1} & I_2 \end{bmatrix} \cdot \begin{bmatrix} A_1 & B \\ B^t & A_2 \end{bmatrix} \cdot \begin{bmatrix} u^1 \\ u^2 \end{bmatrix} = 0 \tag{20}$$

which gives

$$\hat{A}_2 u^2 = 0 \tag{21}$$

with

$$\hat{A}_2 = A_2 - B^t A_1^{-1} B . \tag{22}$$

But, because $\hat{A}_2$ is obtained (cf. [7]) after the Gaussian elimination on the first $n-p$ columns of A, from (21) we have

$$u^2 = 0 \tag{23}$$

which implies (19).

Corollary. In the hypothesis of the theorem 1 and if $D = \mathrm{diag}(A)$ satisfies (16) we havw

$$\min \left\{ \| u - I_p^n u_p \|_0^2 \mid u_p \in R^p \right\} \leq b_1 \cdot \| u \|_1^2 \tag{24}$$

with $b_1 > 0$ given by

$$b_1 = \alpha / \mu_1 . \tag{25}$$

Proof. We have

$$\min \left\{ \| u - I_p^n u_p \|_0^2 \mid u_p \in R^p \right\} \leq \| u \|_0^2 . \tag{26}$$

But, by (23), (19) and (16) we obtain

$$\| u \|_0^2 = \langle Du , u \rangle = \alpha \langle u^1, u^1 \rangle \leq \alpha / \mu_1 \| A_1^{1/2} u^1 \| = (\alpha / \mu_1) \| u \|_1^2 . \tag{27}$$

Then, (26), (27) prove (24).

Theorem 2. If $u \in R^n$ satisfies (11), I_p^n is given by (18) and $D = \mathrm{diag}(A)$ verfies (16) we have

$$\langle A u , u \rangle \geq c_0 \left(\| u^1 \|^2 + \| B u^2 \|^2 \right) \tag{28}$$

with $c_0 > 0$ given by

$$c_0 = \frac{(1 + \alpha \cdot \mu_1) + \sqrt{(1 + \mu_1^2)(1 + \alpha^2)}}{2\alpha} \tag{29}$$

Proof. From (11) we obtain

$$B^t D_1^{-1} R_1 u_1^1 - B^t D_1^{-1} B u^2 + A_2 u^2 = 0 \tag{30}$$

with $D_1 = \mathrm{diag}(A_1) = \alpha \cdot I_1$ and $R_1 = D_1 - A_1$. From (30) it results (for $u^2 \neq 0$

else we can apply the theorem 1)

$$\langle A_2 \, u^2 \, , \, u^2 \rangle \; = \; \langle B^t \, D_1^{-1} B \, u^2 \, , \, u^2 \rangle \; - \; \langle B^t \, D_1^{-1} R_1 \, u^1 \, , \, u^2 \rangle \qquad (31)$$

Then

$$\langle A \, u \, , u \rangle \; = \; \langle A_1 \, u^1 , u^1 \rangle + 2 \langle (I_1 - 1/2 \, D_1^{-1} R_1 \,) \, u^1 , \, y \rangle + \langle D_1^{-1} y, \, y \rangle \qquad (32)$$

with $y \in \mathbb{R}^{n-p}$ given by

$$y \; = \; B \, u^2. \qquad (33)$$

The relation (32) may be written like

$$\langle A \, u \, , \, u \rangle \; = \langle \; \tilde{A} \begin{bmatrix} u^1 \\ y \end{bmatrix} \, , \begin{bmatrix} u^1 \\ y \end{bmatrix} \; \rangle \qquad (34)$$

where $\tilde{A}$ is the symmetric matrix of dimension $2(n-p)$ given by

$$\tilde{A} \; = \; \begin{bmatrix} A_1 & 1/2 \, (I_1 + 1/\alpha \cdot A_1) \\ 1/2 \, (I_1 + 1/\alpha \cdot A_1) & 1/\alpha \cdot I_1 \end{bmatrix}. \qquad (35)$$

Let $\lambda \in \mathbb{R}$ be an eigenvalue of $\tilde{A}$ and $[x \, , \, y]^t \in \mathbb{R}^{n-p} \oplus \mathbb{R}^{n-p}$ the corresponding eigenvector, that is

$$\tilde{A} \begin{bmatrix} x \\ y \end{bmatrix} \; = \; \lambda \begin{bmatrix} x \\ y \end{bmatrix}. \qquad (36)$$

Writting (36) componentwise we obtain

$$\begin{cases} A_1 \, x + 1/2 \, (\, I_1 + 1/\alpha \, A_1 \,) \, y \; = \; \lambda \, x \\ 1/2 \, (\, I_1 + 1/\alpha \, A_1 \,) \, x + 1/\alpha \, y \; = \; y \, . \end{cases} \qquad (37)$$

It can be proved without difficulty that the casses $y = 0$, $x = 0$ or $x \neq 0$ $y \neq 0$ and $\lambda = 1/\alpha$ lead to some contradictions with the initial hypothesis. Thus, from the second equality of (37) we get

$$y = 1/(\, \lambda - 1/\alpha \,) \, 1/2 \, (\, I_1 + 1/\alpha \, A_1 \,) \, x \, . \qquad (38)$$

From (38) and the first equality from (37) we then obtain

$$\left[A_1 + \frac{\alpha}{4(\lambda \alpha \, -1)} \, (\, I_1 + 1/\alpha \, A_1 \,)^2 \right] x \; = \; \lambda \, x \, . \qquad (39)$$

From (39) it results that $x \in \mathbb{R}^{n-p}$, $x \neq 0$ is an eigenvector of A_1 corresponding to an eigenvalue $\mu \in \{\mu_1, \ldots, \mu_{n-p}\}$, that is

$$A_1 x = \mu \cdot x . \tag{40}$$

From (39)-(40) we then obtain

$$\alpha \lambda^2 - \lambda (1 + \mu \alpha) + \mu - \alpha (1 + \mu / \alpha)^2 = 0 \tag{41}$$

which gives us

$$\lambda = \frac{(1 + \mu \alpha) \pm \sqrt{(1 + \mu^2)(1 + \alpha^2)}}{2 \alpha} \tag{42}$$

Now let $\mathscr{S} \subset \mathbb{R}^{n-p} \times \mathbb{R}^{n-p}$ be the subspace of all vectors z of the form $[u^1, B u^2]^t$, $u^1 \in \mathbb{R}^{n-p}$, $u^2 \in \mathbb{R}^p$ such that $u = [u^1, u^2]^t$ satisfies (11). From (33) - (34) we hve that for all $z \in \mathscr{S}$

$$\langle \tilde{A} z, z \rangle = \langle A u, u \rangle . \tag{43}$$

But because A is positive definite, we have that, if $u \neq 0$, $\langle Au, u \rangle > 0$,

$$\langle \tilde{A} z, z \rangle > 0, \ (\forall) \ z \in \mathscr{S} \ , \ z \neq 0 \tag{44}$$

Then λ from (42) is a positive number given by

$$\lambda = \frac{(1 + \mu \alpha) + \sqrt{(1 + \mu^2)(1 + \alpha^2)}}{2 \alpha} . \tag{45}$$

Now we must observe that the expression on the right of (45) (like function on μ) attains its minimal value for $\mu = \mu_1$ which gives c_o from (29). Then, from (44) we obtain

$$\langle \tilde{A} z, z \rangle \geqslant c_o \| z \|^2 \tag{46}$$

which gives us (28).

Corollary. In the hypothesis of the theorem 2 we have

$$\min \left\{ \| u - I_p^n u_p \|_\oplus^2 \ \middle| \ u_p \in \mathbb{R}^p \right\} \leq b_2 \| u \|_1^2 \tag{47}$$

with $b_2 > 0$ given by

$$b_2 = \frac{2 \alpha}{c_o \cdot \min \{1, \alpha^2\}} . \tag{48}$$

Proof. For an arbitrary $u_p \in \mathbb{R}^p$ we have

$$\| u - I_p^n u_p \|_o^2 = \alpha \cdot \left\{ \| u^1 + 1/\alpha \ B u_p \| + \| u^2 - u_p \|^2 \right\} . \qquad (49)$$

Then

$$\min \left\{ \| u - I_p^n u_p \|_o^2 \ \big| \ u_p \in \mathbb{R}^p \right\} \leq \| u - I_p^n u^2 \|_o^2 = \alpha \ (\| u^1 + 1/\alpha \ B u^2 \|^2) \leq$$

$$\leq \alpha \cdot (\| u^1 \| + 1/\alpha \cdot \| B u^2 \|)^2 \leq 2\alpha \cdot \max \left\{ 1, 1/\alpha^2 \right\} (\| u^1 \|^2 + \| B u^2 \|^2)^2 \leq$$

$$\leq \frac{2\alpha}{c_o \cdot \min \left\{ 1, \alpha^2 \right\}} \ \| u \|_1^2$$

which proves (47).

4. Remarks

In this final section we shall make some observations concerning the case of a general I L U decomposition and a general symmetric matrix A (not necessarily positive definite).For a general I L U decomposition we can give the following continuity argument(not even a proof) : we can think that the constants b_1 and b_2 depend continuously on the entries of the interpolation operator.But (17) and (18) are extreme cases of such an I L U decomposition. Then,making an I L U decomposition an A_1 ,i.e.(see [5], [7])

$$\bar{A}_1 = A_1 + R_1 \qquad , R_1 \geqslant 0 \qquad (50)$$

and defining I_p^n by

$$I_p^n = \begin{bmatrix} -\bar{A}_1^{-1} \ B \\ \\ I_2 \end{bmatrix} . \qquad (51)$$

we can hope that we shall obtain a relation like (24) or (47) with a constant b between b_1 and b_2 .This hypothesis is "supported" by the tests presented in [7](see also [4]).

In the case of an arbitrary symmetric matrix A we have already proved ([7]) that using (17) we obtain the convergence of the two-grid method. For the case (18) we can only obtain a relation of the form

$$\langle A u , u \rangle \geqslant - c \cdot \| u \|^2 \qquad (52)$$

if u satisfies (11) with c > 0 not depending on the dimension of A.The relation (52) is similar with (2.2) from [6] .We have also made some remarks on this way,but for an arbitrary I L U decomposition,in the paper[9] following the ideeas from [2] .

Concerning the construction of the coarse grid we can follow the papers [1] and [10] .The number p and the coarse grid points must be choosen such that every node which is not on the coarse grid must have at least one "strong connection" with a coarse grid node.In this way we obtain a better diagonally dominance of the matrix A_1 (that is,a bigger constant μ_1 from (15)) and,therefore,better constants b_1 or b_2 .

References

[1] Brandt, A. : "Algebraic multigrid theory: the symmetric case",Preprint,Weizmann Inst.of Science,Rehovot,Israel,1983 .

[2] Greenbaum,A. : "Analysis of a multigrid method as an iterative technique for solving linear systems",SIAM J. Num. Anal.,vol. 21,nr.3, (1984),pp.473 to 485 .

[3] Hackbusch, W. : "Multigrid methods and applications",Springer-Verlag Berlin(1985) .

[4] Juncu, Gh.,Popa, C. : "Introduction to the multigrid method"(in romanian),Ed. Tehnica,Bucharest(1991) .

[5] Meijerink, J.A.,van der Vorst, H.A. : "An iterative solution method for linear systems of which the coefficient matrix is a symmetric M − matrix",Math. of Comp.,vol.31,nr.137(1977),pp.148 to 162 .

[6] Nicolaides, R.A. : "On multigrid convergence in the indefinite case" Report ICASE − VA,nr. 23655,NASA(1977) .

[7] Popa, C. : "I L U decomposition for coarse grid correction step on algebraic multigrid",in "Multigrid Methods II : Special Topics and Applications,G M D Studien,nr.189(1991) .

[8] Popa, C. : "On the smoothing properties of S O R relaxation for algebraic multigrid method",Studii si Cerc. Matem.,vol.5(1989) .

[9] Popa, C. : "Algebraic multigrid for approximate solutions of symmetric systems,",to appear in Proceed. of the 4-th Conf. on Math. and Applications,Timisoara,november 1991 .

[10] Ruge, J.,Stuben, K. : "Algebraic multigrid (AMG)",Arbeitspapiereder G M D,nr. 210 (1986) .

The Smoothing Property for Regular Splittings

Arnold Reusken

Department of Mathematics and Computing Science
Eindhoven University of Technology
P.O. Box 513, 5600 MB Eindhoven, The Netherlands

Summary

In this paper we discuss convergence of multigrid methods with respect to the *maximum norm* for 2D elliptic boundary value problems. Our analysis uses Hackbusch's framework based on the Smoothing Property and Approximation Property (cf. [4]). We present a rather general framework for establishing the Smoothing Property in the maximum norm. The analysis fits in nicely with the classical theory of diagonally dominant matrices and of M-matrices.

1. Introduction

We consider multigrid methods applied to standard linear finite element discretizations of second order elliptic boundary value problems in 2D. It is well-known that the convergence of these methods can be analyzed using Hackbusch's framework based on the Approximation Property and Smoothing Property (cf. [4]).

In this paper we study convergence with respect to the *maximum norm*. Some first results about multigrid convergence in the maximum norm are given in [8,9]. Below in §4 we briefly discuss the Approximation Property in the maximum norm. A detailed analysis concerning this subject can be found in [9]. The main topic of this paper is to present a rather general framework for establishing the Smoothing Property in the maximum norm. We show connections between nonexpansive splittings (i.e. $A = W - R$ with $\|W^{-1}R\|_\infty \le 1$) and smoothers and between (weak) regular splittings and smoothers.

Our analysis fits in nicely with the classical theory of diagonally dominant matrices and of M-matrices. In particular our analysis can be applied to the ILU-factorization of an M-matrix. Our results differ from the results concerning the Smoothing Property in [4], [11,12] because we use the maximum norm instead of the energy norm or the euclidean norm and we do not need any symmetry conditions.

2. Continuous problem, discretization and two-grid method

Let $\Omega \subset I\!\!R^2$ be a bounded convex polygonal domain. We consider the following second order variational boundary value problem with homogeneous Dirichlet boundary conditions (and $f \in L^2(\Omega)$):

$$
\begin{cases}
\text{find } \varphi^* \in H_0^1(\Omega) \text{ such that for all } \psi \in H_0^1(\Omega) : \\[2mm]
\displaystyle\sum_{|\alpha|,|\beta|\leq 1} \int_\Omega a_{\alpha\beta}(x)\,(D^\alpha\varphi^*(x))\,(D^\beta\psi(x))\,dx = \int_\Omega f(x)\,\psi(x)\,dx \;.
\end{cases}
\tag{2.1}
$$

We use the following notation

$$
a(\varphi,\psi) = \sum_{|\alpha|,|\beta|\leq 1} \int_\Omega a_{\alpha\beta}(x)\,(D^\alpha\varphi)\,(D^\beta\psi)\,dx \quad (\varphi,\psi \in H_0^1(\Omega)) ,
\tag{2.2}
$$

and we assume that the coefficients $a_{\alpha\beta}$ are such that this bilinear form is bounded and elliptic on $H_0^1(\Omega)$.

We use standard linear finite element discretizations on a sequence of nested quasi-uniform triangulations $\{\mathcal{T}_k \mid k \in I\!\!N_0\}$. This results in a sequence of nested finite dimensional function spaces

$$
\Phi_0 \subset \Phi_1 \subset \ldots \subset \Phi_k \subset \ldots \subset H_0^1(\Omega) ,
$$

with

$$
\Phi_k := \{\varphi \in C(\bar\Omega) \mid \varphi \text{ is linear on every } T \in \mathcal{T}_k , \;\; \varphi = 0 \text{ on } \partial\Omega\} .
$$

With $\mathcal{T}_k$ (or Φ_k) there corresponds a mesh size parameter h_k. The collection of interior grid points in $\mathcal{T}_k$ is denoted by $\{x_k^i\}_{i\in J_k}$ for some indexset J_k with $\#J_k =: n_k = O(h_k^{-2})$. We use the notation $U_k = I\!\!R^{n_k}$. The standard basis in Φ_k is given by the functions $\varphi_k^i \in \Phi_k$ which satisfy $\varphi_k^i(x_k^j) = \delta_{ij}$ $(i,j \in J_k)$. This induces the natural bijection

$$
P_k : U_k \to \Phi_k , \quad P_k(u) = \sum_{i\in J_k} u_i\varphi_k^i .
\tag{2.3}
$$

On U_k we use a scaled euclidean inner product:

$$
< u,v >_k = h_k^2 \sum_{i\in J_k} u_i v_i .
\tag{2.4}
$$

Below adjoints are always defined with respect to this inner product on U_k and the L^2-inner product on Φ_k. The maximum norm on U_k is denoted by $\|\cdot\|_\infty$. The norms $\|\cdot\|_\infty$ (on U_k) and $\|\cdot\|_{L^\infty}$ (on Φ_k) induce associated operator norms which are denoted by $\|\cdot\|_\infty$.

Galerkin discretization results in a stiffness matrix $L_k : U_k \to U_k$ defined by

$$
< L_k u, v >_k = a(P_k u, P_k v) \quad \text{for all} \quad u,v \in U_k .
\tag{2.5}
$$

131

In the two-grid (and multigrid) method we use a prolongation $p = p_k : U_{k-1} \to U_k$ and restriction $r = r_k : U_k \to U_{k-1}$ defined by

$$p = P_k^{-1} P_{k-1} , \quad r = p^* . \tag{2.6}$$

The iteration matrix of the smoothing method (cf. §3) is denoted by S_k. The standard two-grid method, with ν pre-smoothing iterations, for solving a system $L_k u_k = b_k$ then has the following iteration matrix:

$$T_k(\nu) = (I - pL_{k-1}^{-1} r L_k) S_k^{\nu} = (L_k^{-1} - pL_{k-1}^{-1} r) L_k S_k^{\nu} . \tag{2.7}$$

In §3 we give a detailed analysis of the following Smoothing Property (cf. [4]):

$$\|L_k S_k^{\nu}\|_{\infty} \leq \xi_0(\nu) h_k^{-2} \quad \text{with} \quad \xi_0(\nu) \to 0 \quad \text{for} \quad \nu \to \infty . \tag{2.8}$$

In §4 we briefly discuss the following Approximation Property:

$$\|L_k^{-1} - pL_{k-1}^{-1} r\|_{\infty} \leq C \, |\ln h_k|^2 \, h_k^2 . \tag{2.9}$$

Combination of (2.8) and (2.9) yields a bound for the two-grid contraction number with respect to the maximum norm:

$$\|T_k(\nu)\|_{\infty} \leq \xi_0(\nu) \, |\ln h_k|^2 . \tag{2.10}$$

3. The Smoothing Property

The usual technique for proving the Smoothing Property requires symmetry or a nearly symmetric situation, and yields results in the energy norm or in the euclidean norm. We refer to Hackbusch [4] and Wittum [11,12] where smoothing and the construction of smoothers are discussed in a general framework.
In this section we present a rather general framework for establishing the Smoothing Property with respect to the *maximum norm*. Our analysis is based on a new technique, that was first used in [7], which does not require symmetry. One of the main results of this section (cf. Corollary 3.6 and Criterion 3.9) is a connection between weak regular splittings and smoothing methods. We note that our analysis fits in nicely with the theory of diagonally dominant matrices and of M-matrices.

We recall that the approach to the Smoothing Property used by Hackbusch [4] and Wittum [11,12] is based on the following elementary lemma:

> *If A is symmetric and $\sigma(A) \subset [0,1]$ then*
> $$\|A(I - A)^{\nu}\|_2 \leq \eta_0(\nu) := \frac{\nu^{\nu}}{(\nu + 1)^{\nu+1}} \quad (\sim \frac{1}{e\nu} \; \text{for} \; \nu \to \infty) .$$

Also our analysis below is based on an elementary lemma (cf. Lemma 3.2) where instead of the condition "A is symmetric and $\sigma(A) \subset [0,1]$" we use the condition "$\|A\|_{\infty} \leq 1$".
The following function turns out to be important for our analysis:

$$\xi_0(\nu) := 2^{1-\nu} \binom{\nu}{[\frac{1}{2}\nu]} \qquad (\nu \in I\!N_0) \ . \tag{3.1}$$

Lemma 3.1. *The following holds for $\nu \geq 1$:*

$$\sqrt{2} \leq \sqrt{\nu}\, \xi_0(\nu) < 2\sqrt{\frac{2}{\pi}} \quad \text{if } \nu \text{ is even} \tag{3.2.a}$$

$$\sqrt{2} \leq \sqrt{\nu+1}\, \xi_0(\nu) < 2\sqrt{\frac{2}{\pi}} \quad \text{if } \nu \text{ is uneven} \tag{3.2.b}$$

$$\lim_{\nu \to \infty} \sqrt{\nu}\, \xi_0(\nu) = 2\sqrt{\frac{2}{\pi}} \ . \tag{3.2.c}$$

Proof. Define the following sequences for $k \in I\!N_0$:

$$a_k = \binom{2k}{k} \sqrt{k}\, 2^{-2k} \ , \quad b_k = \int_0^{\frac{\pi}{2}} \sin^k(x)\, dx \ . \tag{3.3}$$

Elementary analysis yields that $(b_k)_{k \geq 0}$ is monotonically decreasing and

$$b_k = \frac{k-1}{k}\, b_{k-2} \ , \quad b_{2k+1} = (2^k\, k!)^2 / (2k+1)! \ , \quad b_{2k} = \tfrac{1}{2}\pi (2k)! / (2^k\, k!)^2 \ .$$

From this it follows that for $k \geq 1$

$$1 < \frac{b_{2k}}{b_{2k+1}} = \frac{2k+1}{2k}\, \frac{b_{2k}}{b_{2k-1}} < 1 + \frac{1}{2k} \ ,$$

and thus $\displaystyle\lim_{k \to \infty} \frac{b_{2k}}{b_{2k+1}} = 1$. Using $\frac{b_{2k}}{b_{2k+1}} = \pi(1 + \frac{1}{2k})\, a_k^2$ for $k \geq 1$ yields

$$\lim_{k \to \infty} a_k = \frac{1}{\sqrt{\pi}} \ . \tag{3.4}$$

Furthermore $(a_k)_{k \geq 0}$ is monotonically increasing, so

$$\tfrac{1}{2} \leq a_k < \frac{1}{\sqrt{\pi}} \quad \text{for all} \quad k \geq 1 \ . \tag{3.5}$$

Also

$$\sqrt{\nu}\, \xi_0(\nu) = 2\sqrt{2}\, a_{\frac{1}{2}\nu} \quad \text{if} \quad \nu \text{ is even} \tag{3.6}$$

and

$$\sqrt{\nu+1}\, \xi_0(\nu) = 4\sqrt{\nu+1}\, 2^{-(1+\nu)} \tfrac{1}{2} \binom{\nu+1}{\frac{1}{2}(\nu+1)} = 2\sqrt{2}\, a_{\frac{1}{2}(\nu+1)} \quad \text{if} \quad \nu \text{ is uneven} \ . \tag{3.7}$$

From (3.5), (3.6), (3.7) the results in (3.2.a–c) easily follow. $\qquad\square$

Lemma 3.2. *Let B be a square matrix with $\|B\|_\infty \leq 1$. Then the following holds for $\nu \in \mathbb{N}_0$:*

$$\|(I - B)\,(\tfrac{1}{2}(I+B))^\nu\|_\infty \leq \xi_0(\nu) . \tag{3.8}$$

Proof. Note that

$$(I - B)\,(I + B)^\nu = (I - B) \sum_{k=0}^{\nu} \binom{\nu}{k} B^k = I - B^{\nu+1} + \sum_{k=1}^{\nu} \left(\binom{\nu}{k} - \binom{\nu}{k-1} \right) B^k .$$

So

$$\|(I - B)\,(I + B)^\nu\|_\infty \leq 2 + \sum_{k=1}^{\nu} \left| \binom{\nu}{k} - \binom{\nu}{k-1} \right| .$$

Using $\binom{\nu}{k} \geq \binom{\nu}{k-1} \Leftrightarrow k \leq \tfrac{1}{2}(\nu+1)$ and $\binom{\nu}{k} = \binom{\nu}{\nu-k}$ we get

$$\sum_{k=1}^{\nu} \left| \binom{\nu}{k} - \binom{\nu}{k-1} \right| =$$

$$= \sum_{1}^{[\frac{1}{2}(\nu+1)]} \left(\binom{\nu}{k} - \binom{\nu}{k-1} \right) + \sum_{[\frac{1}{2}(\nu+1)]+1}^{\nu} \left(\binom{\nu}{k-1} - \binom{\nu}{k} \right)$$

$$= \sum_{1}^{[\frac{1}{2}\nu]} \left(\binom{\nu}{k} - \binom{\nu}{k-1} \right) + \sum_{m=1}^{[\frac{1}{2}\nu]} \left(\binom{\nu}{m} - \binom{\nu}{m-1} \right)$$

$$= 2 \sum_{k=1}^{[\frac{1}{2}\nu]} \left(\binom{\nu}{k} - \binom{\nu}{k-1} \right) = 2 \left(\binom{\nu}{[\frac{1}{2}\nu]} - \binom{\nu}{0} \right) .$$

So $\|(I - B)\,(\tfrac{1}{2}(I+B))^\nu\|_\infty \leq 2^{-\nu}(2 + 2(\binom{\nu}{[\frac{1}{2}\nu]} - 1)) = \xi_0(\nu).$ $\qquad\square$

Remark 3.3. The estimate in (3.8) is sharp: Let B be the following $n \times n$ matrix

$$B = \begin{bmatrix} 0 & 1 & & \emptyset \\ & 0 & \ddots & \\ & & \ddots & \ddots \\ \emptyset & & \ddots & 1 \\ & & & 0 \end{bmatrix},$$

then for $\nu \leq n - 2$ equality holds in (3.8).

We also note that the analysis above holds for other norms too (cf. [7]).

Definition 3.4. $A = W - R$ with W regular is called a *nonexpansive splitting* if $\|W^{-1}R\|_\infty \leq 1$ holds.

In the remainder we only consider splittings $A = W - R$ for which W is regular.

Theorem 3.5. *Let* $A = W - R$ *be a nonexpansive splitting. Then the following holds for* $\theta \in]0, \frac{1}{2}]$:

$$\|A(I - \theta W^{-1}A)^\nu\|_\infty \leq \frac{1}{2\theta}\,\xi_0(\nu)\,\|W\|_\infty \quad (\nu \in I\!N_0)\,.$$

Proof. Let $B := I - 2\theta W^{-1}A = 2\theta W^{-1}R + (1 - 2\theta)\,I$. Then $\|B\|_\infty \leq 1$ and using Lemma 3.2 we get

$$\|A(I - \theta W^{-1}A)^\nu\|_\infty = \|\frac{1}{2\theta}\,W(I - B)\,(\tfrac{1}{2}(I + B))^\nu\|_\infty \leq \frac{1}{2\theta}\,\xi_0(\nu)\,\|W\|_\infty\,. \qquad \square$$

Corollary 3.6. If we apply Theorem 3.5 to the stiffness matrices $L_k(k \geq 0)$ of §2 this yields the following. Let $L_k = W_k - R_k$ be such that for every k:

(a)   we have a nonexpansive splitting; (3.9.a)

(b) $\|W_k\|_\infty \leq Ch_k^{-2}$ with C independent of k. (3.9.b)

Then for $\theta \in]0, \frac{1}{2}]$ the following Smoothing Property holds:

$$\|L_k(I - \theta W_k^{-1}L_k)^\nu\|_\infty \leq \frac{C}{2\theta}\,\xi_0(\nu)\,h_k^{-2}\,.$$

Usually for the splittings used as a smoother in multigrid methods the condition (3.9.b) is fulfilled. The condition (3.9.a) is more severe. Below we give some criteria for nonexpansive splittings.

First we recall some definitions. A matrix $A = (a_{ij})$ is called *weakly diagonally dominant* if $\sum_{j \neq i} |a_{ij}| \leq |a_{ii}|$ for all i. A splitting $A = W - R$ is called a *regular splitting* if $W^{-1} \geq 0$ and $R \geq 0$ hold (with "$\geq$" entrywise ordering); $A = W - R$ is called a *weak regular splitting* if $W^{-1} \geq 0$ and $W^{-1}R \geq 0$.

Variants of Criterion 3.7 below are well-known. In the literature one can find variants with somewhat stronger assumptions (e.g. A irreducibly diagonally dominant or strictly diagonally dominant, cf. [10], [13]) which are then sufficient to yield convergence, i.e. $\rho(W^{-1}R) < 1$. Our assumptions do not imply convergence but are sufficient for $\|W^{-1}R\|_\infty \leq 1$ to hold.

Criterion 3.7. *Consider a system* $Au = b$ *with* A *an* $n \times n$ *nonsingular and weakly diagonally dominant matrix. We write* $A = D - L - U$ *with* $D = \mathrm{diag}(A)$, L *strictly lower triangular,* U *strictly upper triangular and consider the following splittings:*

(a) $\qquad W = D$, $\quad R = L + U \quad$ *(Jacobi)* ;

(b) $\qquad W = \dfrac{1}{\omega} D - L$, $\quad R = \left(\dfrac{1}{\omega} - 1\right) D + U \quad with \quad \omega \in \,]0,1] \quad$ *(SOR; Gauss-Seidel).*

Then $A = W - R$ is a nonexpansive splitting.

Proof. The proof for the Jacobi method is trivial (note that W is regular because A is regular and weakly diagonally dominant).

For the splitting in (b) the proof runs as follows. Take $v \in I\!\!R^n$ with $\|v\|_\infty = 1$, and define $z := W^{-1}Rv$.

From $(\frac{1}{\omega} D - L) z = ((\frac{1}{\omega} - 1) D + U) v$ it follows that for $1 \le k \le n$

$$z_k = (1 - \omega)\, v_k + \omega\, \frac{1}{a_{kk}} \left(-\sum_{j=1}^{k-1} a_{kj} z_j + \sum_{j=k+1}^{n} a_{kj} v_j \right) .$$

From this it is easy to prove with induction that for $1 \le i \le n$ the following holds:

$$|z_i| \le (1 - \omega) + \omega\, \frac{1}{|a_{ii}|} \sum_{j \neq i} |a_{ij}| \le 1 \, ,$$

and thus $\|z\|_\infty = \|W^{-1}Rv\|_\infty \le 1$. $\qquad\qquad\qquad\square$

Below we use the notation $e = (1, 1, \ldots, 1)^T$.

Criterion 3.8. *Let $A = W - R$ be a splitting such that $W^{-1}R \ge 0$. Then $A = W - R$ is a nonexpansive splitting iff $W^{-1}Ae \ge 0$.*

Proof. Because $W^{-1}R \ge 0$ we have that $\|W^{-1}R\|_\infty = \max_i (W^{-1}Re)_i$. And thus

$$\|W^{-1}R\|_\infty \le 1 \quad \Longleftrightarrow \quad \max_i ((I - W^{-1}A)\,e)_i \le 1$$

$$\Longleftrightarrow \quad 1 - (W^{-1}Ae)_i \le 1 \quad \text{for all} \quad i$$

$$\Longleftrightarrow \quad (W^{-1}Ae)_i \ge 0 \quad \text{for all} \quad i$$

$$\Longleftrightarrow \quad W^{-1}Ae \ge 0 \, . \qquad\qquad\qquad\square$$

An important application of Criterion 3.8 is given in Criterion 3.9 below. The latter criterion relates weak regular splittings with nonexpansive splittings (and thus with smoothing methods).

Criterion 3.9. *Let A be such that $Ae \ge 0$. Then every weak regular splitting $A = W - R$ is a nonexpansive splitting.*

Proof. Because $A = W - R$ is a weak regular splitting we have $W^{-1}R \ge 0$ and $W^{-1} \ge 0$. Combined with $Ae \ge 0$ this also yields $W^{-1}Ae \ge 0$. Application of Criterion 3.8 proves that $A = W - R$ is a nonexpansive splitting. $\qquad\qquad\qquad\square$

Remark 3.10. Consider $L_k = W_k - R_k$ and assume that this is a regular splitting for every k.

In [12] Wittum proves the following: If for every k L_k is symmetric, W_k is symmetric positive definite and $\|W_k\|_2 \leq Ch_k^{-2}$ holds then, if we use a damping factor $\theta \in\,]0,1[$, the Smoothing Property holds in the euclidean norm. Criterion 3.9 and Corollary 3.6 yield the following: If for every k $L_k e \geq 0$ and $\|W_k\|_\infty \leq Ch_k^{-2}$ hold then, if we use a damping factor $\theta \in\,]0,\frac{1}{2}]$, the Smoothing Property holds in the maximum norm.

We give a few examples in which Criterion 3.9 applies.
We assume that A is such that $a_{ij} \leq 0$ for all $i \neq j$ and $a_{ii} > 0$ for all i and that A is irreducibly diagonally dominant (these conditions are easy to verify in practice). These assumptions imply that A is an M-matrix and also that $Ae \geq 0$ holds.
Note that, due to the diagonal dominance, results for the Jacobi and Gauss-Seidel splittings follow already from Criterion 3.7.
As a first important application we consider an ILU factorization of A. It is shown in [5] that the corresponding splitting $A = W - R$ is regular. Criterion 3.9 now yields that this is a non-expansive splitting.
Other examples are block versions of Jacobi, Gauss-Seidel or ILU. Suitable block versions correspond to regular splittings (cf. [10], [1]) and Criterion 3.9 then can be used to prove that these are nonexpansive splittings.

4. Approximation Property

For a multigrid convergence analysis the results of §3 should be combined with an analysis of the Approximation Property (cf. [4]). In this paper we do not present such an analysis, however, for completeness we give some results from the paper [9] in which a detailed analysis of the Approximation Property with respect to the maximum norm is given.

We outline a main result from [9] (cf. [9] for details). Assume that $\partial\Omega$ and the coefficients in the differential operator (cf. §2) are sufficiently smooth and that the principal part of this differential operator is symmetric (i.e. $a_{\alpha\beta} = a_{\beta\alpha}$ if $|\alpha| = |\beta| = 1$). Assume that the triangulations, which are quasi-uniform (cf. §2), satisfy $h_k h_{k+1}^{-1} \leq c$ with c independent of k. Then the following holds:
There are constants k_0 and C such that for all $k \geq k_0$:

$$\|L_k^{-1} - pL_{k-1}^{-1}r\|_\infty \leq Ch_k^2 |\ln h_k|^2 . \tag{4.1}$$

Remark 4.1. We briefly comment on the proof of the estimate (4.1). Two important arguments in the proof are the following. Firstly, we use that the mass matrix $P_k^* P_k$ has a condition number which is $O(1)$ $(h_k \downarrow 0)$ with respect to the maximum norm. This is a result due to Descloux [2]. Secondly, we use the following finite element asymptotic error estimate due to Rannacher and Frehse [3,6] (with $f \in L^\infty(\Omega)$ and φ_k^* the Galerkin approximation of φ^* in Φ_k):

$$\|\varphi_k^* - \varphi^*\|_{L^\infty} \leq Ch_k^2 |\ln h_k|^2 \|f\|_{L^\infty} .$$

Combination of the results in §3 with the Approximation Property (4.1) immediately yields an (asymptotic) estimate for the contraction number of the two-grid iteration matrix (cf. §2):

$$\|T_k(\nu)\|_\infty \leq C \, \frac{1}{\sqrt{\nu}} \, |\ln h_k|^2 \, . \tag{4.2}$$

So instead of an "optimal" bound $C\nu^{-1}$ for the contraction number in the energy norm (or the euclidean norm) we obtain a "nearly optimal" bound $C\nu^{-\frac{1}{2}} |\ln h_k|^2$ if we use the maximum norm. Furthermore the bound in (4.2) is sharp in a certain sense: For a concrete (very regular) example it is shown in [9] that the contraction number with respect to the maximum norm of a standard two-grid method with a fixed number of pre-smoothing iterations is bounded from below by $C |\ln h_k|$.

References

[1] O. AXELSSON, S. BRINKKEMPER, V.P. IL'IN, *On some versions of incomplete block-matrix factorization iterative methods*, Linear Algebra Appl., 58 (1984), pp. 3–15.

[2] J. DESCLOUX, *On finite element matrices*, SIAM J. Numer. Anal., 9 (1972), pp. 260–265.

[3] J. FREHSE, R. RANNACHER, *Eine L^1-Fehlerabschätzung für diskrete Grundlösungen in der Methode der finiten Elemente*, Tagungsband "Finite Elemente" Bonn. Math. Schr. 1976.

[4] W. HACKBUSCH, *Multi-grid Methods and Applications*, Springer, Berlin, 1985.

[5] J.A. MEIJERINK, H.A. VAN DER VORST, *An iterative solution method for linear systems of which the coefficient matrix is a symmetric M-matrix*, Math. Comp., 31 (1977), pp. 148–162.

[6] R. RANNACHER, *Zur L^∞-Konvergenz linearer finiter Elemente beim Dirichlet-problem*, Math. Z., 149 (1976), pp. 69–77.

[7] A. REUSKEN, *A new lemma in multigrid convergence theory*, RANA Report 91–07, Department of Mathematics and Computing Science, Eindhoven University of Technology, 1991.

[8] A. REUSKEN, *On maximum norm convergence of multigrid methods for two-point boundary value problems*, to appear in SIAM J. Numer. Anal.

[9] A. REUSKEN, *On maximum norm convergence of multigrid methods for elliptic boundary value problems*, submitted.

[10] R.S. VARGA, *Matrix Iterative Analysis*, Prentice-Hall, Englewood Cliffs, 1962.

[11] G. WITTUM, *On the robustness of ILU-smoothing*, SIAM J. Sci. Stat. Comput., 10 (1989), pp. 699–717.

[12] G. WITTUM, *Linear iterations as smoothers in multigrid methods: Theory with applications to incomplete decompositions*, Impact of Computing in Science and Engineering, 1 (1989), pp. 180–215.

[13] D.M. YOUNG, *Iterative solution of large linear systems*, Academic Press, New York, 1971.

ON THE STABILITY OF THE ILU – METHOD
FOR SINGULAR PERTURBED
FINITE ELEMENT PROBLEMS

Stefan Sauter

Institut für Informatik und praktische Mathematik
Olshausenstr. 40   W – 2300 Kiel, Germany

Abstract: The ILU–method for solving systems of linear equations arising by discretizing partial differential equations via finite elements is known as a robust smoother in a multigrid procedure. The stability of the ILU–method is the fundamental property in the theoretical examination of this algorithm. The usual (sufficient) conditions to ensure stability are maximum angles conditions for the grid, which are easy to satisfy for model problems, but are quite unrealistic for irregular domains or anisotropic problems. In our paper we analyse an elliptic problem on a degenerate grid and prove, that the stability does not depend on the perturbation. Numerical tests will support the theoretical examination.

1 Examples of perturbed problems

1.1 Problem of anisotropy

For $\epsilon \in \mathbb{R}_+$, consider the elliptic partial differential equation:

$$\frac{1}{1 + \epsilon^2} \cdot \operatorname{div} \begin{bmatrix} 1 + \epsilon^2 & -2 \\ -2 & 4 \end{bmatrix} \operatorname{grad} u = g \ \text{ in } \ \Omega := (0,1)^2 \tag{1.1}$$

with homogeneous Dirichlet boundary data: $u|_{\partial\Omega} := 0.$

As ϵ tends to zero the positive definite matrix $w(\epsilon) := \begin{bmatrix} 1 + \epsilon^2 & -2 \\ -2 & 4 \end{bmatrix}$ becomes singular, the problem degenerates in the direction of the eigenvector: $e_0 = \begin{bmatrix} 2 \\ 1 \end{bmatrix}$.

If we discretize this problem with linear finite elements using a square grid triangulation, consisting of pairs of triangles of the form ◪ , and number the grid points lexicographically, the arising stiffness matrix K^ϵ can be characterized by the stencil:

$$K^\epsilon := \frac{1}{\epsilon} \cdot \begin{bmatrix} & -1 & -1 \\ \frac{1 - \epsilon^2}{2} & 3 + \epsilon^2 & \frac{1 - \epsilon^2}{2} \\ -1 & -1 & \end{bmatrix}, \tag{1.2}$$

which is positive definite for all $\epsilon \in \mathbb{R}_+$ and becomes singular as ϵ tends to zero.

This was an example, where we have a singular perturbed equation on a regular grid. In the following example we show that problem 1.1 is equivalent to a regular equation on a perturbed grid.

1.2 Regular equation on a perturbed grid

Let us transform the variables (x, y) of equation (1.1) into (x', y') using the transformation:

$$\begin{bmatrix} x' \\ y' \end{bmatrix} := \frac{1}{2} \begin{bmatrix} 2 & 1 \\ 0 & \epsilon \end{bmatrix} \begin{bmatrix} x \\ y \end{bmatrix}.$$

Then in the new coordinates equation (1.1) takes the following form:

$$\Delta u = g \quad \text{in } \Omega^\epsilon$$

$$u = 0 \quad \text{on } \partial\Omega^\epsilon,$$

where the quadrangle Ω_ϵ has the corner points $(0,0)$, $(1,0)$, $(\frac{1}{2}, \frac{\epsilon}{2})$, $(\frac{3}{2}, \frac{\epsilon}{2})$. The transformation of the square grid triangulation of example 1.1 results in the triangulation depicted in figure 1. The corresponding stiffness matrix yields the same stiffness matrix K^ϵ as in example 1.1.

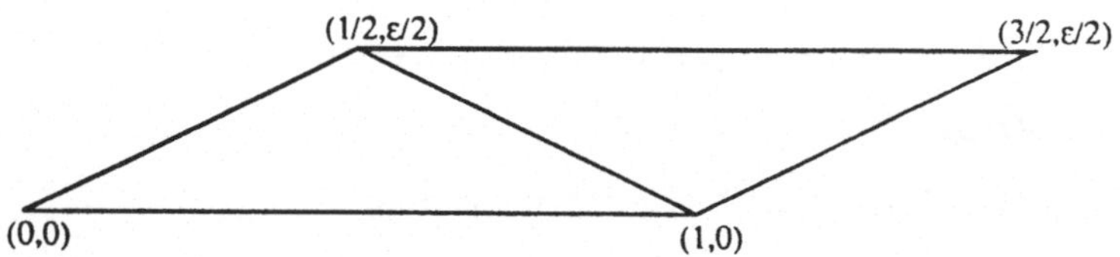

fig. 1 domain Ω^ϵ with corners $(0, 0)$, $(1, 0)$, $(1/2, \epsilon/2)$, $(3/2, \epsilon/2)$.

The next example shows, that such degenerate triangles are typical if one has to resolve complicate boundary structures, by using refinement strategies, which refine coarse grids, projecting new boundary points onto the physical surface.

1.3 Problem of complicated boundary structure

In the field of Limnologie it is well known, that lakes form systems having eigenmodes and eigenfrequencies. These phenomena cause the internal seiches of the lakes and are responsable for the plankton distribution. They are therefore of important interest for the fishermen and observed from of old [2, 12]. The underlying equations are the shallow water equations, which can be simplified in the case of lake Constance(Bodensee/Germany) to the scalar partial differential eigenvalue problem [1]:

$$\text{div}\,(g \cdot h\,\text{grad})\,u + \omega^2 u = 0 \quad \text{in } \Omega \ \text{(surface of the lake)}$$

$$\frac{\partial}{\partial n} u = 0 \qquad \text{on } \partial\Omega \ \text{(shore of the lake)} ,$$

$$\text{(1.3)}$$

where $h = h(x, y)$ denotes the distance between the calm surface and the bottom of the lake, g the constant of gravity. u denotes the vertical displacement of the oscillation relative to the calm surface, ω the corresponding eigenvalue.
In this problem the main difficulty arises from the complicated microstructure of the boundary (in our example about approx. 170 reentrant corners).

We must use high grid point densities near the boundary to resolve the topological structure of the lake, but also in the interior the grid must not be too poor, in order to permit an approximation of eigenmodes with higher oscillations.

The fastest procedure to solve this problem is the multigrid grid algorithm for eigenvalue problems as introduced by Hackbusch [4, 5]. To do this procedure efficiently we have to use a rather coarse coarse–grid, and refine this grid succesively, adapting the new grid to the physical boundary of the lake.

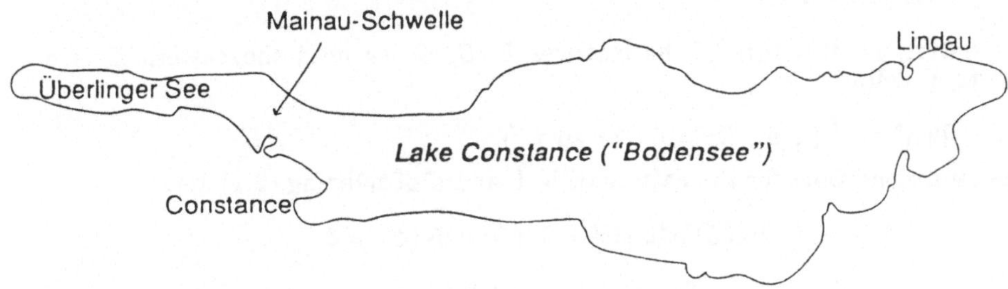

fig 2: shape of lake Constance

In the case of such a complicated boundary it seems unavoidable to get deformed triangles. Figure 2 shows the coarse grid triangulation and the following sequence of pictures zooms out typical difficulties, which arise by refining the grids.

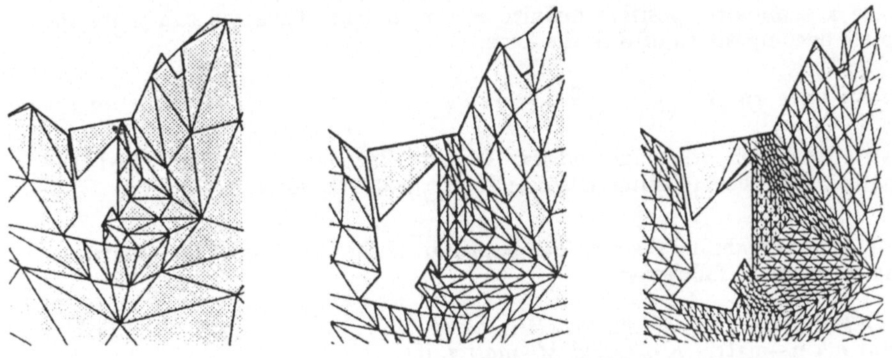

fig. 3 magnification of grids of different refinement stages around the harbour of Lindau

These examples show the importance of analysing the behaviour of numerical methods for perturbed grids. For the ILU–method this will be done in the next section.

2 The stability of the ILU–method for a perturbed grid

The idea of the incomplete factorization is, to perform the usual $L\ U$ decomposition of a matrix K only on a prescribed pattern and to set the other elements to zero during the elimination process.This results in a splitting of the form:

$$K = L\ U - C \qquad (2.1)$$

with the rest–matrix C and the "sparse" lower (upper) triangular matrix L (U). The corresponding iteration scheme to solve the equation $K\,x = b$ then proceeds as follows:

arbitrary starting guess: x_0

$$\text{iteration step:} \qquad x_{i+1} := x_i - M^{-1}(K\,x_i - b),$$

$$\text{(2.2)}$$

with $M := L\,U$.

To control the structure of the matrices L, U, C, we need the pattern $\mathfrak{P}$ of a matrix K, defined by:

$$\mathfrak{P}(K) := \{\ (i,\,j)\ ;\ K_{i,j} \neq 0\ ,\ \text{for all } i,\,j\}\ .$$

The usual conditions for the pattern of L, U and C of splitting (2.1) are:

$$\mathfrak{P}\,(L) \cap \mathfrak{P}\,(C) = \mathfrak{P}\,(U) \cap \mathfrak{P}\,(C) = \emptyset,$$

$$\mathfrak{P}\,(K) = \mathfrak{P}\,(L) \cup \mathfrak{P}\,(U)\ . \qquad (2.3)$$

The fundamental step in the numerical examination of the algorithm is to prove the numerical stability of the method, that is the investigation of the method for $h \to 0$.

DEFINITION 2.1: *numerical stability of the ILU method*
Let K be a symmetric positive definite $n \times n$ matrix. Then we can write the incomplete decomposition of K in the form:

$$K = (L + D)\,D^{-1}\,(L + D)^{\mathrm{T}} - C \qquad (2.4)$$

with strictly lower triagonal matrix L and diagonal D. C is called the rest–matrix. The decomposition is *stable*, if $D_{i,i} > 0$ for all i .

A general stability theorem has been proved for the class of M – matrices. For this we need the following:

DEFINITION 2.2 : *M–matrix, H–matrix*
A regular $n \times n$ –matrix K is called *M–matrix*, if
$$K_{i,i} > 0 \quad \text{for all } i, \qquad\qquad K_{i,j} \leq 0 \quad \text{for all } i \neq j,$$

$$(K^{-1})_{i,j} \geq 0 \quad \text{for all } i,j\,.$$

A regular $n \times n$ –matrix K is an *H–matrix*, if the matrix $\breve{K}$, defined by:
$$\breve{K}_{i,i} := K_{i,i} \qquad\qquad\qquad \breve{K}_{i,j} := -\,|K_{i,j}| \qquad \forall\, i \neq j,$$

is an M–matrix.

THEOREM 2.3 Let K be a symmetric positive definite M – resp. H–matrix. Then the ILU–decompostion of K is stable.

PROOF: see [6], [7].
The following lemma shows the difficulties in using the above theorem to ensure the stability of the ILU–method for matrices arising from finite element discretizations on general grids:

Lemma 2.4: Let τ_0 be a triangulation of a polygonal domain Ω. Let τ_1 be generated by combining the midpoints in each triangle of τ_0. If we discretize Poisson's equation with continous, piecewise linear finite elements over τ_1 the arising stiffness matrix K_{τ_1} has the following property:

K_{τ_1} is an M–matrix $\Leftrightarrow$ no triangle of τ_0 contains an angle larger than $\frac{\pi}{2}$.

Proof: Let e be an edge of a triangulation τ not lying on the boundary of τ, and Δ_1^e, Δ_2^e be the two triangles having e as an edge. For $i = 1,2$ let α_i^e be the angle in Δ_i^e, lying opposite to the side e. If no triangle of τ_0 contains an angle larger than $\frac{\pi}{2}$, then τ_1 does not contain an edge e with $\alpha_1^e + \alpha_2^e > \pi$. In the opposite case, there is such an edge. The lemma follows from the fact, that K_τ is an M–matrix, iff τ does not contain an edge e with $\alpha_1^e + \alpha_2^e > \pi$. The proof for this is given in [8, theorem 6.8.7].

QED

Obviously the matrix K^ϵ of example 1.1 resp. example 1.2 is not in the class of M–matrices for $\epsilon < 1$. Test–calculations show, that the matrices K^ϵ are not in the class of H–matrices too for a wide range of parameters ϵ. We investigate now the behaviour of the diagonal elements $D_{i,i}^\epsilon$ of the splitting of K^ϵ according (2.4) as ϵ tends to zero. The surprising result will be, that the stability in our example does not depend on ϵ, meaning in the words of example 1.2, that it does not depend on degenerating triangles.

Theorem 2.5 : The ILU–method as defined in (2.3), (2.4) for the matrix K^ϵ of example 1.1 and 1.2 is stable for all $\epsilon > 0$. For the elements $D_{i,i}^\epsilon$ – formula (2.4) – the estimate :

$$K_{i,i}^\epsilon \geq D_{i,i}^\epsilon \geq 0.612 \cdot K_{i,i}^\epsilon \tag{2.5}$$

holds for $\epsilon \leq 1$, for all i, and is unrelated to the dimension of K^ϵ.

Proof: For $\epsilon \geq 1$, K^ϵ is an M–matrix, thus the stability of the ILU–method follows from theorem (2.3). For $\epsilon < 1$ the proof is rather technical.

The idea is, to prove, using a nested induction, that the elements of L^ϵ can be estimated by some constant times $K^\epsilon_{i,i}$ up to an index m (e.g. $L^\epsilon_{m,m-1}$); with this the diagonals $D^\epsilon_{i,i}$ can be estimated for $m < i \leq m + k$ and we can proceed to estimate $L^\epsilon_{m+k,m+k-1}$. We omit here the details. The complete proof can be found in [10, theorem 2.1.1] .

QED

3 The ILU–method as a smoother in a multigrid process

In order to use the ILU–iteration in a multigrid process, we assume that we discretize a partial differential operator K of order $2 \cdot m$ on a hierarchy of grids τ_l, $1 \leq l \leq l_{max}$, yielding the sequence of equations:

$$K_l\, x_l = b_l . \tag{3.1}$$

For a detailed introduction into the theory of multigrid methods we refer to the book of Hackbusch [5]. Following his convergence analysis, one has to prove, that the "smoother" satisfies the so–called smoothing property:

DEFINITION 3.1 : *smoothing property*
Let K_l be split into:

$$K_l = M_l - N_l , \qquad M_l \text{ invertible.}$$

The corresponding linear iterative scheme then reads:

$$x_{i+1} = S_l\, x_i + M_l^{-1} b , \tag{3.2}$$

with
$$S_l := I_l - M_l^{-1} K_l .$$

The iterative method has the smoothing property for problem (3.1), if

$$\| K_l \cdot S_l^\nu \| \leq C \cdot h_l^{-2m} \cdot \eta\,(\nu) \tag{3.3}$$

holds true with $\eta(\nu) \longrightarrow 0$ for $\nu \longrightarrow \infty$, where $\|\cdot\|$ denotes the spectral norm and h_l the stepsize.

To use the ILU–scheme as a smoother in a multigrid process one should modify the smoother in a suitable way. A common modification of the ILU–method is the ILU_β – scheme as proposed by Gustafson[3]. Instead of condition (2.3): $C_{i,i} = 0$, one generalizes this condition to:

$$C_{i,i} = \beta \cdot \sum_{j \neq i} C_{i,j} .$$

This modification has the disadvantage, that the stability of ILU_β is not ensured, even if the usual ILU–decomposition is stable as can be seen from the following example.

EXAMPLE 3.2 : Let the symmetric and positive definite matrix K be defined by:

$$K := \begin{bmatrix} 1 & -\frac{1}{3} & 0 & \frac{1}{2} \\ -\frac{1}{3} & \frac{1}{6} & 0 & 0 \\ 0 & 0 & 1 & 0 \\ \frac{1}{2} & 0 & 0 & 1 \end{bmatrix} .$$

The ILU–decomposition of K takes the following form:

$$K := \begin{bmatrix} 1 & 0 & 0 & 0 \\ -\frac{1}{3} & \frac{1}{3\sqrt{2}} & 0 & 0 \\ 0 & 0 & 1 & 0 \\ \frac{1}{2} & 0 & 0 & \sqrt{\frac{3}{4}} \end{bmatrix} \cdot \begin{bmatrix} 1 & -\frac{1}{3} & 0 & \frac{1}{2} \\ 0 & \frac{1}{3\sqrt{2}} & 0 & 0 \\ 0 & 0 & 1 & 0 \\ 0 & 0 & 0 & \sqrt{\frac{3}{4}} \end{bmatrix} - \begin{bmatrix} 0 & 0 & 0 & 0 \\ 0 & 0 & 0 & -\frac{1}{6} \\ 0 & 0 & 0 & 0 \\ 0 & -\frac{1}{6} & 0 & 0 \end{bmatrix} .$$

Now it is easy to calculate that ILU_β does not exist for $\beta = \frac{1}{3}$ and at least one of the elements of the diagonal–matrix D of ILU_β becomes negative for all $\beta > \frac{1}{3}$.

The following variant of the ILU–method is due to Wittum [13]. One has to perform the standard ILU–iteration, to get the splitting (2.4). The diagonal matrix $\mathfrak{N} := \text{diag} \{ \, n_{i,i} , 1 \leq i \leq n \}$ is defined by:

$$n_{i,i} := \max \{ \nu_i, \underline{\nu} \} \quad \text{for } 1 \leq i \leq n ,$$

with

$$\nu_i := \sum_{j=1}^{n} |n_{i,j}|$$

and

$$\underline{\nu} := \{ \, \nu_i , \nu_i > 0, 1 \leq i \leq n \} .$$

Now we can define the modified diagonal matrix D_β by: $D_\beta := D + \beta \cdot \mathfrak{N}$ and M_β via:

$$M_\beta := (L + D_\beta) \, D_\beta^{-1} \, (L + D_\beta)^{\mathrm{T}} , \tag{3.4}$$

and the iteration proceeds according (3.2). For this modification the stability property follows immediately from the stability of standard ILU–scheme. Wittum proved in [13] the smoothing property for this modification:

THEOREM 3.3: Let K_l be a symmetric and positive definite differential operator of order $2 \cdot m$. Let K_l be symmetric and positive definite and the ILU–splitting (2.4) be stable. Let the matrix M_l satisfy:

$$\| M_l \| \leq C_M \cdot h_l^{-2m} .$$

Then there exists $\beta_0 \in \mathbb{R}$ so that the ILU–iteration defined by (3.4) satisfies the smoothing property for all $\beta \geq \beta_0$.

4 Numerical examples

To check the quantitative behaviour of the above theory we have performed numerical experiments for the matrix of example 1.1 resp. 1.2. The convergence behaviour of the examined algorithm is measured by the convergence rate κ defined by the average of the contraction number κ_i of the residuals r_i:

$$\kappa_i := \| r_i \| \, / \, \| r_{i-1} \| \; .$$

As pointed out above the modified ILU–algorithm as a smoother in a multigrid process shows excellent convergence properties, which seems to be independent of the singular perturbation, if we choose $\beta := 0.8$ (cf. figure 4).

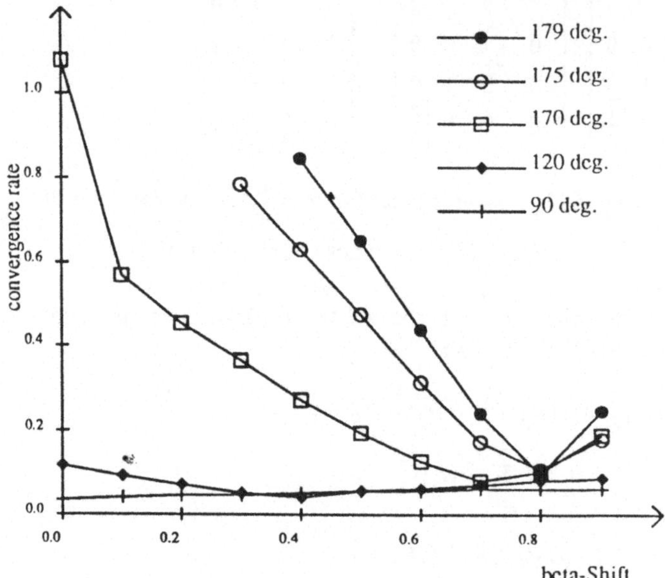

fig. 4 Convergence rates of ILU_β in a multigrid algorithm, using a V–cycle with two pre– and two post–smoothing steps. We used seven grids; the finest contains 4096 nodal points. $\alpha_{\max}$ denotes the largest angle of τ_l.

What we have theoretically proved was only, that for each perturbation parameter ϵ, we can choose a suitable β, to get grid–independent convergence rates. The numerical tests indicate however that the behaviour of $\mathrm{ILU}_{0.8}$ is robust, that means the convergence is independent of the perturbation parameter. Finally we perform numerical calculations to solve problem 1.3. To get significant convergence rates, we fix ω in 1.3 and add the following function f to the righthand side of the equation:

$$f(x, y) := \sin(x/5) \cdot \sin(y/3) - c,$$

where c is chosen in such a way that $\int_\Omega f(x, y) \, dx \, dy = 0$. That means we compute a regular Neumann problem. The finest grid consists of 2080 triangles and 1129 nodal points. The convergence rates were about 0.125, although the grid contains rather obtuse angles.

To solve the eigenvalue problem, we use the nested multigrid algorithm for eigenvalue–problems as introduced by Hackbusch in [5]. An outline of the theory focussing on the proof of the smoothing property for the modified ILU–smoother can be found in Sauter/Wittum [11]. Algorithmic details are described in [9] where also a vectorizing variant is presented. For our example we use a coarse grid, containing 88 nodal points. Thus we could compute the first 11 eigenmodes observing variing convergence rates. A plot of all eigenfunctions and a comparison with experimental measurement can be found in [8]. As an example we plot the shape of the 10.th eigenmode.

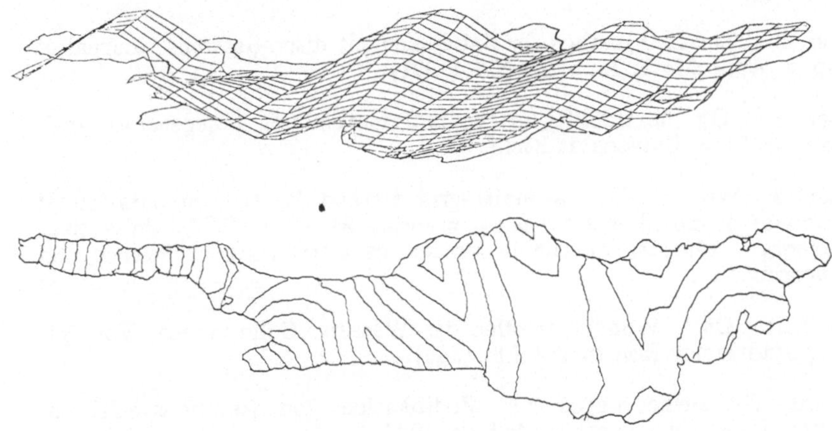

Fig. 5. Shape and level lines of the 10th eigenmode; $\lambda_{10} = 11.65$ min.

REFERENCES

[1] Bäuerle, E.: Die Eigenschwingungen abgeschlossener, zweigeschichteter Wasserbecken bei variabler Bodentopographie. Dissertation, Institut für Meereskunde, Kiel, 1981.

[2] Forel, F. A.: Die Schwankungen des Bodensees. Schr. Verein. Gesch. Bodensees, 22, 49–77 (1893).

[3] Gustafsson, I.: A class of first order factorization methods. BIT 18, 142–156, 1978.

[4] Hackbusch, W.: On the convergence of approximate eigenvalues and eigenfunctions of elliptic operators by means of a multi–grid method. SIAM J. Numer. Anal., 16, (1979).

[5]   Hackbusch, W.: Multi–grid methods and applications. Springer, Berlin, Heidelberg (1985).

[6]   Manteuffel, T. A.: An incomplete factorization technique for positive definite linear systems. Math. of Comp., Vol. 34, 1980, p. 473–497.

[7]   Meijerink, J. A. , van der Vorst, H. A.: An iterative solution for linear systems of which the coefficient matrix is a symmetric M–matrix. Math. of Comp. 31, 148–162, (1977).

[8]   Sauter, S.: Ein Mehrgitterverfahren zur Berechnung der Eigenschwingungen abgeschlossener Wasserbecken. Master thesis, Universität Heidelberg, 1989.

[9]   Sauter, S.: The ILU method for finite–element discretizations. Journal of Comp. a. Appl. Math. 36, p. 91–106, (1991).

[10]   Sauter, S.: On the stability of the ILU method for a degenerate grid. Report Nr. 9113, Universität Kiel (1991).

[11]   Sauter, S., Wittum, G.: A multi–grid method for the computation of eigenmodes of closed water basins, preprint 91 – 15, IWR, Universität Heidelberg , 1991, to appear in Impact of Computing in Science and Engineering.

[12]   Schulthaiss, Ch.: Wunder anloffen des Wassers. Collectaneen, Vol. VI, 80–81, Stadtarchiv Konstanz A I 8 (1549).

[13]   Wittum, G.: Bemerkungen zur Modifikation von Iterationsverfahren. Preprint, IWR, Universität Heidelberg, 1992.

ILU as a solver in a parallel multi-grid flow prediction code

M. Schäfer and E. Schreck
Lehrstuhl für Strömungsmechanik, Universität Erlangen-Nürnberg
Cauerstr. 4, D-8520 Erlangen, Germany

Summary

In this paper ILU methods are considered as solvers in a parallel multigrid flow prediction code. A parallelization strategy for the inherently recursive ILU methods is discussed with respect to their numerical and parallel efficiency. By a number of numerical experiments the computational performance of the ILU approach is investigated. The studies include comparisons with preconditioned conjugate gradient methods as well as investigations concerning the influence of flow properties and algorithm parameters on the convergence behaviour of the methods.

1. Introduction

The major part of the computing time when solving flow problems is spent for the solution of sparse linear systems. Therefore, in order to have competitive flow prediction codes, it is of high importance to use efficient linear system solvers. It is well known that ILU methods belong to the most efficient methods for this task, especially in connection with multigrid techniques.

In this paper we investigate the performance of ILU methods when used as solvers in the iterative solution process of a parallel flow prediction code. The underlying numerical method for the fluid flow computations is a finite volume method with colocated arrangement of variables on non-orthogonal blockstructured grids. For the pressure velocity coupling an iterative algorithm of SIMPLE type with underrelaxation is used and a multigrid method, in which the SIMPLE method acts as the smoother, is implemented for convergence acceleration (e.g. Hortmann et al. [5]). The code is parallelized by means of a grid partitioning technique with non-overlapping subdomains for the use on local memory multiple instruction multiple data (MIMD) machines. The parallelization strategy is directly related to the concept of blockstructured grids.

The ILU method we consider, is a parallelized version of the strongly implicit procedure of Stone [11], which comprises the standard ILU method (e.g. Axelsson-Barker [2]). For comparison preconditioned conjugate methods (PCG) with basic iterative method preconditioning (e.g. Adams [1]) are also considered, where as basic iterative method a parallel version of a symmetric block Gauß-Seidel method (e.g. Schäfer [10]) is used.

By a number of numerical experiments we investigate the performance of the considered methods with respect to their computational efficiency. The studies include comparisons of ILU algorithms and PCG methods, comparisons of different block coupling strategies and investigations concerning the influence of the processor topology and characteristic flow parameters like the Reynolds number on the convergence behaviour of the methods.

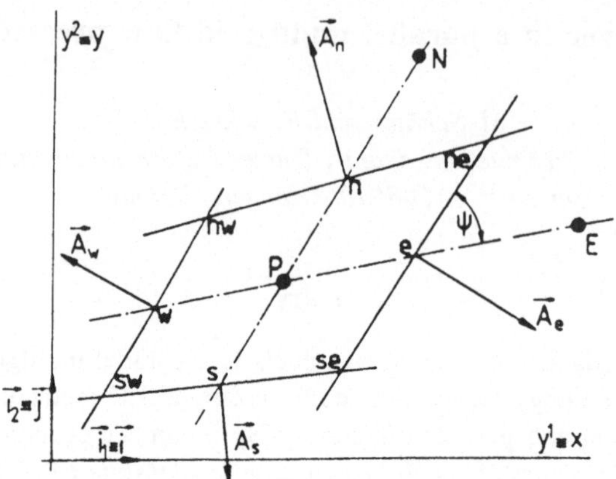

Figure 1. *A typical control volume and labelling scheme.*

2. Numerical method and parallel implementation

The fluid dynamic problems considered for this study are two-dimensional (plane or axisymmetric), steady, laminar flows. These problems are described by the conservation equations for the transport of mass, momentum, and heat, which read (in cylindrical coordinates):

$$\frac{\partial}{\partial x}(\rho U) + \frac{\partial}{r\partial r}(\rho r V) = 0 \tag{1}$$

$$\frac{\partial}{\partial x}\left(\rho UU - 2\mu\frac{\partial U}{\partial x}\right) + \frac{\partial}{r\partial r}\left(\rho r VU - \mu r\left(\frac{\partial U}{\partial r} + \frac{\partial V}{\partial x}\right)\right) = -\frac{\partial P}{\partial x} + \rho g_x \tag{2}$$

$$\frac{\partial}{\partial x}\left(\rho UV - \mu\left(\frac{\partial U}{\partial r} + \frac{\partial V}{\partial x}\right)\right) + \frac{\partial}{r\partial r}\left(\rho r VV - 2r\mu\frac{\partial V}{\partial r}\right) = -\frac{\partial P}{\partial r} - 2\mu\frac{V}{r^2} + \rho g_r \tag{3}$$

$$\frac{\partial}{\partial x}\left(\rho UT - \frac{\mu}{Pr}\frac{\partial T}{\partial x}\right) + \frac{\partial}{r\partial r}\left(\rho r VT - r\frac{\mu}{Pr}\frac{\partial T}{\partial r}\right) = 0 \;. \tag{4}$$

Here, ρ is the density, U and V are the axial and radial velocity components a, x and r are the axial and radial coordinate directions, T is the temperature, Pr is the Prandtl-number, μ is the dynamic viscosity, and g_i is the component of the gravitational acceleration vector in the corresponding coordinate direction. In the case of plane flows, the axial and radial directions x and r correspond to the Cartesian coordinates x and y, respectively; in addition, r is replaced by unity in the above equations, the underlined terms in (3) are omitted and (4) is not used. In case of swirling axi-symmetric flows the derivatives in the azimuthal direction are neglected, which renders the problem being two-dimensional.

The solution domain is discretized in quadriliteral (in general non-orthogonal) finite volume cells. The transport equations are integrated over the control volumes (CV), leading to a balance equation for the fluxes through the CV faces, F_{nb}, where nb=e, w, n, s, and the volumetric sources, S, cf. Fig. 1. Thus,

$$F_e - F_w + F_n - F_s = S \;.$$

The convection and diffusion contribution to the flux F are evaluated using a central differencing scheme. This leads for each CV to an algebraic equation of the form:

$$A_p \Phi_p + \sum_{nb} A_{nb} \Phi_{nb} = S_\Phi \ , nb = E,W,N,S$$

where Φ stands for U, V, W or T. For the solution domain as a whole, a matrix equation result. Solution methods for such systems are discussed in detail in the next section.

The solution of the coupled set of nonlinear equations for U, V, W, T, and P is based on the SIMPLE algorithm [7]. The discretized momentum equations are solved using values for the pressure, mass fluxes and temperature from the previous iteration. With the resulting velocity field the pressure correction equation is assembled and solved. The mass fluxes, velocities, and pressure are then corrected with the evaluated pressure correction and afterwards, in the case of a non-isothermal flow, the temperature equation is solved. This procedure is repeated until a convergence criterion is reached. For this it is required that the maximum sum of the absolute residuals in all equations is reduced to a prescribed value. Details about the discretization and the pressure-velocity coupling can be found in Perić et al. [8] and Demirdžić and Perić [3].

The iterative solution procedure described above removes efficiently only those Fourier components of the error whose wavelengths are comparable to the grid spacing. For this reason, the number of iterations increases linearly with increasing number of grid points, resulting in a quadratic increase in computing time. To reduce this increase a multigrid (MG) scheme is applied. It is implemented in a so called full multigrid fashion, where the SIMPLE method acts as a smoother. The MG procedure is described in detail in Hortmann et al. [5].

For the treatment of complex geometries the code allows the use of blockstructured non-orthogonal grids. This (geometrical) blockstructuring approach is also well suited as a base for the parallelization of the algorithm by domain decomposition, since each block can be assigned to individual processors. For handling the block coupling and to reduce the data communication between the blocks auxiliary control volumes along the block boundaries containing the corresponding boundary values of the neighboring block are introduced. During the iterative algorithm the boundary data of neighboring blocks have to be interchanged from time to time. In addition to this local data transfers some global communication is required to transfer residuals for convergence check. The flow chart of the underlying parallel SIMPLE method is sketched in Fig. 2. For the block coupling we have investigated two strategies. In the first approach the block boundary data are exchanged after each iteration of the solver as indicated in Fig. 2 (SIMPLE1). In the second one the exchange within the solver is simply omitted (SIMPLE2). Clearly, the second variant will yield better results with respect to the parallel efficiency, but due to the strong decoupling the numerical efficiency will be poorer (e.g. Perić et al. [9]).

3. Linear System Solvers

In order to obtain an efficient algorithm for fluid flow computations on a parallel computer, the crucial point in the method described in the preceding section is the choice of the numerical method for the solution of the sparse linear systems that arise during the iterative solution process. The parallelization of the other components of the methods (assembly of

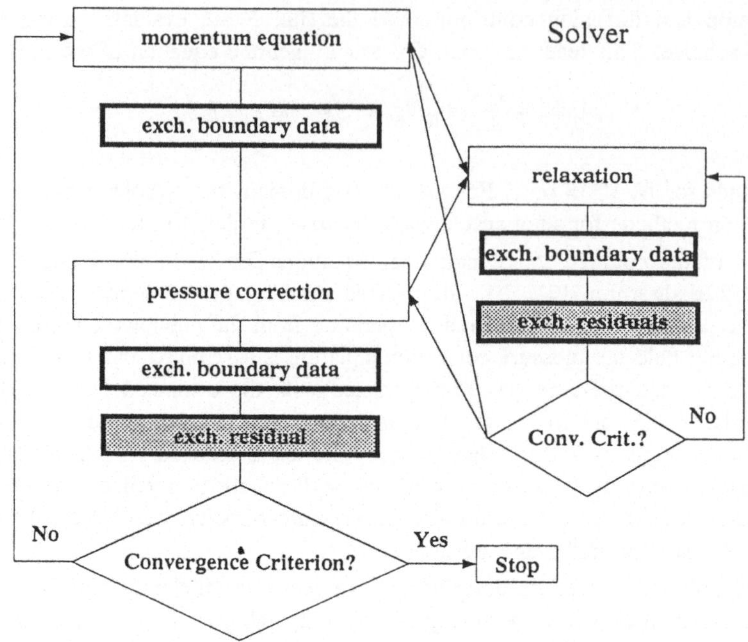

Figure 2: *Control flow in the parallel SIMPLE algorithm with boundary exchanges.*

the systems, prolongation, restriction, etc.), with the considered concept of blockstructured grids and domain decomposition, is straightforward.

In this paper we consider parallel variants of incomplete factorization methods (e.g. Wesseling [12]) and preconditioned conjugate gradient methods (Meijerink and van der Vorst [6]) which in its sequential versions both rank among the most efficient methods to deal with such problems.

Let us denote by

$$Mx = r \tag{5}$$

the system that we want to solve. Numbering the unknowns in a natural way according to our blockstructure, i.e. lexicographically within each block and block by block, the coefficient matrix M gets a blockstructure of the form

$$
M = \begin{pmatrix}
B_{11} & B_{12} & \cdot & \cdot & & \cdot & & B_{1m} \\
B_{21} & B_{22} & \cdot & & & & & \cdot \\
\cdot & & \cdot & \cdot & & & & \cdot \\
\cdot & & & \cdot & \cdot & & & \cdot \\
\cdot & & & & \cdot & \cdot & & B_{m-1,m} \\
B_{m1} & \cdot & & \cdot & \cdot & B_{m,m-1} & B_{mm}
\end{pmatrix},
$$

where m is the number of blocks (i.e. processors). The matrices B_{ii}, $i = 1, \ldots m$ on the diagonal have the usual pentadiagonal structure resulting from a five point discretization

152

in one block. The other matrices B_{ij}, $i, j = 1, \ldots m$, $i \neq j$ represent the coupling of the blocks. They have nonzero entries (only in the main diagonal) only if block i is neighbored to block j, otherwise all entries vanish. According to this structure we split M into a local part M_L and a coupling part M_C:

$$M = M_L + M_C,$$

with

$$M_L = \begin{pmatrix} B_{11} & & & & \\ & B_{22} & & 0 & \\ & & \ddots & & \\ & 0 & & \ddots & \\ & & & & B_{mm} \end{pmatrix}, \quad M_C = \begin{pmatrix} 0 & B_{12} & \cdot & \cdot & & B_{1m} \\ B_{21} & 0 & \cdot & & & \cdot \\ \cdot & & \cdot & \cdot & & \cdot \\ \cdot & & & \cdot & & \cdot \\ \cdot & & & \cdot & 0 & B_{m-1,m} \\ B_{m1} & \cdot & \cdot & \cdot & B_{m,m-1} & 0 \end{pmatrix}.$$

In general, ILU factorization methods are defined by an iteration process of the form

$$y_{n+1} = y_n - H^{-1}(My^n - r) \tag{6}$$

with some incomplete decomposition $H = LU$ of M into upper and lower triangular matrices L and U respectively (e.g. Axelsson-Barker [2]). Our parallelization approach for these methods consists in an incomplete decomposition of M_L instead of M:

$$H = \begin{pmatrix} L_1 U_1 & & & & \\ & L_2 U_2 & & 0 & \\ & & \ddots & & \\ & 0 & & \ddots & \\ & & & & L_m U_m \end{pmatrix}.$$

For the local decompositions $L_i U_i$, $i = 1, \ldots m$ in each block we apply the strongly implicit procedure of Stone [11], which is defined by the recurrence formulas:

$$h_W^{ij} = b_W^{ij} / \left(1 + \alpha h_N^{i-1,j}\right),$$
$$h_S^{ij} = b_S^{ij} / \left(1 + \alpha h_E^{i-1,j}\right),$$
$$h_P^{ij} = b_P^{ij} + \alpha \left(h_W^{ij} h_N^{i-1,j} + h_S^{ij} h_E^{i,j-1}\right) - h_W^{ij} h_E^{i-1,j} - h_S^{ij} h_N^{i,j-1},$$
$$h_N^{ij} = \left(b_N^{ij} - \alpha h_W^{ij} h_N^{i-1,j}\right) / h_P^{ij},$$
$$h_E^{ij} = \left(b_E^{ij} - \alpha h_S^{ij} h_E^{i,j-1}\right) / h_P^{ij},$$

where $b_W^{ij}, \ldots, b_E^{ij}$ and $h_W^{ij}, \ldots, h_E^{ij}$ denote the coefficients of a block of M_L and H, respectively, with indication according to the stencil shown in Figure 3. The iteration parameter α has to be in the interval $[0, 1)$. For $\alpha = 0$ the standard ILU approach (e.g. Axelsson-Barker [2]) results. The influence of α on the convergence will be discussed in the next section.

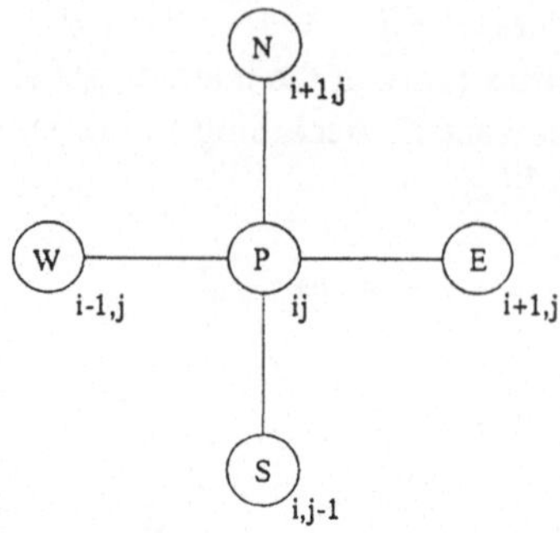

Figure 3: Five point stencil with indices.

The communication requirements for the above ILU approach consists in one exchange of the block boundaries in each iteration. The updated values are needed for the matrix vector multiplication in (6) with the coupling matrix M_C.

The basic idea of PCG methods is to transform (5) into an equivalent system

$$P^{-1}Mx = P^{-1}r \qquad (7)$$

with a nonsingular preconditioning matrix P, and to apply a conjugate gradient type method to (7). For our comparison we restrict ourselves to the solution of the symmetric positive definite pressure correction equations (the solution of these equations takes the major part of the computing time), such that the classical conjugate gradient of Hestenes and Stiefel [4] can be used.

The rate of convergence of PCG methods is determined by the condition number of the matrix $P^{-1}M$ (e.g. Axelsson-Barker [2]): the smaller the condition number, the more rapid is the convergence. Therefore, on the one hand P should be an approximation to M in some sense and on the other hand, since required in each step of a PCG method, the inversion of P should be as simple as possible. Clearly, a compromise between the two extreme cases $P = M$ and $P = Id$ (identity matrix) has to be found.

We concentrate here on a basic iterative method preconditioning with a 1-step symmetric block Gauß-Seidel method (SGS1). Details concerning this approach can be found in Schäfer [10].

Like the ILU approach the basic CG method requires one exchange of block boundaries in each iteration. In addition to that, two global communication steps per iteration are necessary for computing two scalar products. The SGS1 preconditioning in each iteration require one further exchange of block boundaries.

4. Numerical results

For our numerical investigations we consider the following two test problems (see Perić et al. [9]): a pipe flow with obstacle and a natural convection flow in a square cavity. For all computations a Meiko Computing Surface with T800 transputers is used.

At first the influence of the parameter α on the numerical behaviour of the considered ILU approach under various situations is studied. From theory α should be in the interval $(0,1)$ (Stone [11]). For $\alpha = 0$ the standard ILU method results. The performance of the method increases with increasing α until an optimal value α_{opt} is reached. Increasing α beyond this value results first in a rapid decrease in efficiency and finally the method diverges. Figure 4 show the computing times versus α for the pipe flow (256×64 finest grid, 4 multigrid levels) and the natural convection flow (128×128 finest grid, 4 multigrid levels) with different numbers of processors (i.e. subdomains). For the pipe flow a stripe decomposition on a processor ring is employed whereas for the natural convection problem a box decomposition together with a processor torus is used. All cases show the behaviour described above. For both problems one can observe that with increasing number of processors α_{opt} becomes larger and that the optimum becomes less pronounced. This means that with increasing number of processors the method becomes more unsensitive against α variations. This is due to the fact that with increasing number of subdomains the ILU approach becomes closer to a line (stripe decomposition) or point (box decomposition) iterative method such that the variation of α below α_{opt} only has a small influence on the convergence behaviour. For the recirculating natural convection flow α_{opt} is smaller than for the pipe flow, which has a predominant flow direction.

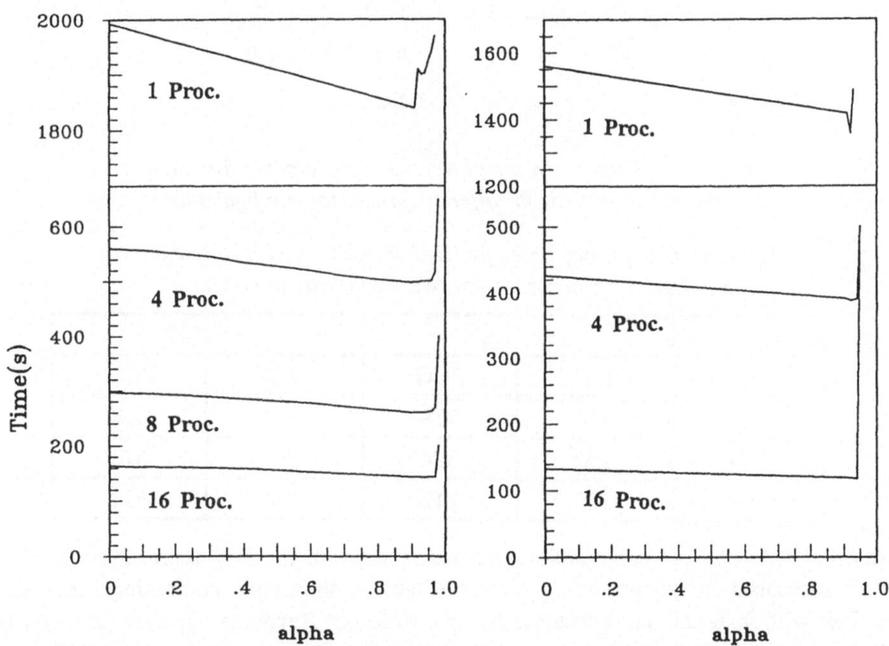

Figure 4: *Computing times versus α parameter for pipe flow (left) and natural convection flow (right) with different numbers of processors.*

From Fig. 5 one can see the influence of the type of domain decomposition together with the α parameter on the convergence behaviour. It shows the computing times versus α for the natural convection flow with the box decomposition 4×4 and the stripe decompositions 16×1 and 1×16. The box partitioning results in the smallest computing time, since in this case the length of the inner boundaries of the subdomains is smaller resulting in less decoupling and therefore better performance.

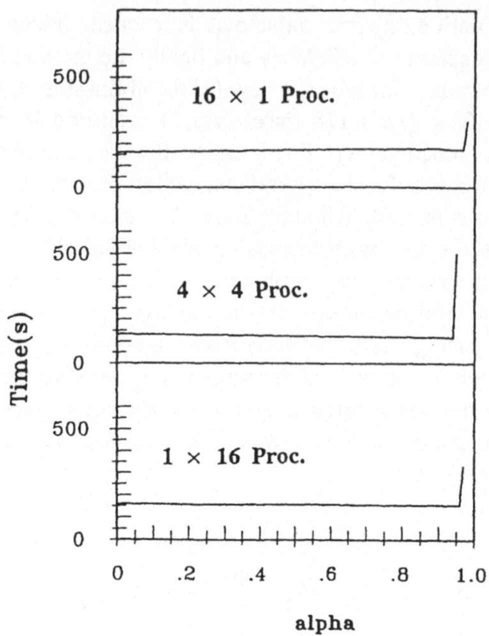

Figure 5: *Computing times versus α parameter for natural convection flow with different processor configurations.*

Table 1: *Computing times for ILU and PCG with different numbers of processors for two Reynolds numbers.*

# Proc.	Re=100		Re=500	
	ILU	PCG	ILU	PCG
4	376	739	553	1296
8	198	384	289	670
16	109	212	160	358

In an other series of test computations we have compared the ILU approach with the PCG method discussed in the preceding section. Table 1 shows the computing times for the pipe flow with 4, 8 and 16 processors for two different Reynolds numbers (256×64 finest grid, 4 multigrid levels). In all cases the ILU approach is superior to the PCG method. Although the PCG method requires less iterations for convergence in all cases, the lower requirements in arithmetic and communication per iteration for the ILU method leads to a shorter comuting time. With increasing Reynolds number the ratio between the computing time with PCG and those with ILU becomes larger. With increasing number of processors this ratio only slightly changes.

In order to study the robustness of our ILU approach we consider the pipe flow with different Reynolds numbers. Figure 6 shows the computing times versus the Reynolds number for different numbers of processors. As one expects, the computing time slightly increases with increasing Reynolds number due to the poorer conditioned linear systems. This increase is rarely affected by the number of processors. The tendency of these results for the pipe flow have been confirmed by various other numerical experiments for different flow situations.

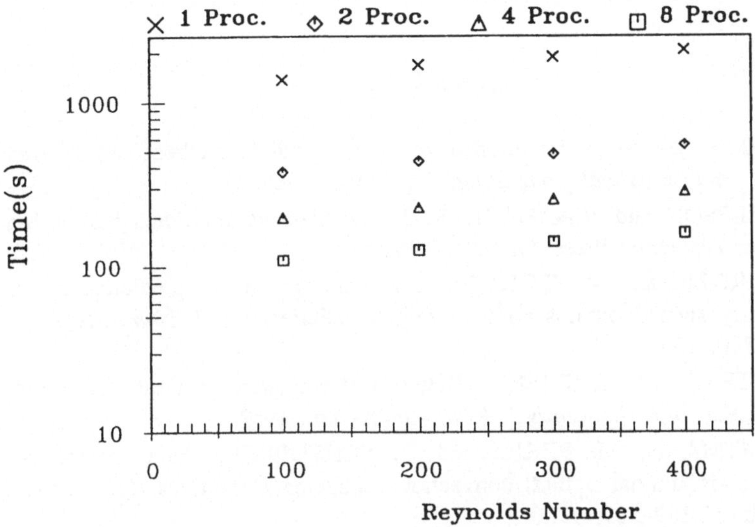

Figure 6: *Computing times versus the Reynolds number for different numbers of processors.*

Finally the influence of the two coupling strategies discussed in Section 2 will be discussed. For this we consider again the pipe flow with a 256×64 finest grid, 4 multigrid levels and Reynolds number 500. A comparison of the computing times for the two coupling modes with different numbers of processors is given in Table 2. The strong decoupling within the EO mode results in a rapid increase of the number of required fine grid iterations if the number of processors is increased. The saving in computing time due to the lower communication requirements is too small to compensate this effect, such that the EI mode becomes more and more superior as the number of processors is increased. Only on architectures with, compared to arithmetic, very slow communication facilities the EO approach may be advantageous. An interesting fact which can also be drawn from Table 2 is that with the EI coupling the efficiency of the present multigrid approach is rarely affected by the decoupling due to the domain decomposition. Looking at Fig. 6 one can observe that this effect is also almost independent on the Reynolds number.

Table 2: *Computing times for the two coupling modes with different numbers of processors.*

# Proc.	SIMPLE1		SIMPLE2	
	Iterations	Time (s)	Iterations	Time (s)
4	73	554	73	699
8	73	291	81	458
16	73	162	129	337

Acknowledgements

The authors thank the Deutsche Forschungsgemeinschaft for the financial support of this research within the Programme "Flow Simulation with Supercomputers" and the Sonderforschungsbereich "Multiprocessor and Network Configurations".

References

[1] L. ADAMS. M-step preconditioned conjugate gradient methods. *SIAM Journal on Scientific and Statistical Computing*, 6:452–463, 1985.

[2] O. AXELSSON and V.A. BARKER. *Finite Element Solution of Boundary Value Problems*. Academic Press, Orlando, 1984.

[3] I. DEMIRDŽIĆ and M. PERIĆ. Finite volume method for prediction of fluid flow in arbitrary shapeddomains with moving boundaries. *Int. J. Num. Meth. in Fluids*, 10:771–790, 1990.

[4] M. HESTENES and E. STIEFEL. Methods of conjugate gradients for solving linear systems. *Nat. Bur. Standards J. Res.*, 49:409–436, 1952.

[5] M. HORTMANN, M. PERIĆ, and G. SCHEUERER. Finite volume multigrid prediction of laminar natural convection: Benchmark solutions. *Int. J. Num. Meth. in Fluids*, 11:189–207, 1990.

[6] J. MEIJERINK and H. VAN DER VORST. An iterative solution method for linear systems of which the coefficient matrix is a symmetric M-matrix. *Mathematics of Computation*, 31:148–162, 1977.

[7] S. V. PATANKAR and D. B. SPALDING. A calculation procedure for heat, mass and momentum transfer in three dimesional parabolic flows. *Int. J. Heat Mass Transfer*, 15:1787–1806, 1972.

[8] M. PERIĆ, R. KESSLER, and G. SCHEUERER. Comparison of finie-volume numerical methods with staggered and colocated grids. *Computers & Fluids*, 16:389–403, 1988.

[9] M. PERIĆ, M. SCHÄFER, and E. SCHRECK. Computation of fluid flow with a parallel multi-grid solver. In *Parallel Computational Fluid Dynamics 91*, pages 297–312, Amsterdam, 1992. Elsevier.

[10] M. SCHÄFER. Numerical solution of the time-dependent axisymmetric Boussinesq equations on processor arrays. *SIAM Journal on Scientific and Statistical Computing*, 1992. to appear.

[11] H. STONE. Iterative solution of implicit approximations of multi-dimensional partial differential equations. *SIAM Journal on Numerical Analysis*, 5:530–558, 1968.

[12] P. WESSELING. Theoretical and practical aspects of a multigrid method. *SIAM Journal on Scientific and Statistical Computing*, 3:387–407, 1982.

New Estimates of the Contraction Number of V-cycle Multi-Grid with Applications to Anisotropic Equations

Rob Stevenson

Department of Mathematics, University of Utrecht,
Budapestlaan 6, P.O. Box 80.010, 3508 TA Utrecht, The Netherlands.

Abstract

In this paper we refine the V-cycle multi-grid convergence proofs of Hackbusch and Wittum. We obtain a sharper bound for the contraction number. With this new bound we are able to prove robustness of the V-cycle applied to anisotropic equations when a suitable smoother is used. For a model problem we give some quantitative results.

1   Introduction

In his well-known multi-grid book ([Hac85]), Hackbusch proved convergence of the V-cycle multi-grid method (MGM). Assumptions that he made are a positive definite discretization, symmetric Galerkin approximations on coarser grids satisfying the so-called *approximation property* and a smoother that is semi-positive definite with respect to the energy scalar product. In order to include ILU-smoothers, Wittum ([Wit89a]) generalized Hackbusch' result to symmetric smoothers with possibly negative eigenvalues. The upper bound for the contraction number that they proved depends on the smoother essentially only by means of $\rho(W_l)$, where W_l is the approximation of the stiffness matrix L_l on level l that defines the smoothing iteration $u_l^{(i+1)} = u_l^{(i)} - W_l^{-1}(L_l u_l^{(i)} - f_l)$. For semi-positive definite smoothers their upper bound is given by $\dfrac{C_A \sup_l \frac{\rho(W_l)}{\rho(L_l)}}{\nu + C_A \sup_l \frac{\rho(W_l)}{\rho(L_l)}}$, where C_A is the constant from the approximation property (cf. [Hac91, (10.6.8)]) and $\nu/2$ is the number of pre- and post-relaxation steps. The *quality* of the smoother has therefore no influence on the estimate of the contraction number. Indeed, among all semi-positive definite smoothers (i.e. $W_l \geq L_l$), as the SGS smoothers, the smoother that gives the smallest bound is the simple damped Jacobi smoother ($W_l = \rho(L_l)I_l$).

In this paper we will refine the proofs of Hackbusch and Wittum. We prove a bound for the contraction number that depends on the smoother essentially by means of (an increasing function of) $\rho(W_l - L_l)$ and thus on the quality of the smoother. For semi-positive definite smoothers, our bound is given by $\dfrac{C_A \sup_l \frac{\rho(W_l - L_l)}{\rho(L_l)}}{\nu + C_A \sup_l \frac{\rho(W_l - L_l)}{\rho(L_l)}}$ which is in all cases better than the original one. Furthermore, as hinted by Hackbusch ([Hac91, Übungsaufgabe 10.7.16(b)]), for all our results we can replace the condition of Galerkin approximations on coarser grids by a somewhat weaker condition. An example of a situation in which this weaker condition is satisfied but not the Galerkin condition is the 5-point discretization on all levels of Poisson's equation on a square with standard 9-point restrictions and prolongations.

A quite different proof of V-cycle convergence was given by Mandel, McCormick and Bank in [MMB87]. Instead of estimating eigenvalues as in our proof, they use projection arguments and norm estimates. As a consequence, their proof is restricted to the situation of Galerkin approximations. At the other hand, their proof has the advantage that it can also be applied to essentially non-symmetric smoothers as the Gauss-Seidel iterations. In case of semi-positive definite smoothers, their bound for the contraction number is the same as ours. For symmetric smoothers with also negative eigenvalues, we did not succeed to derive from their theorems a bound that is as sharp as that of ours. Moreover, their estimate of the contraction number of the W-cycle is equal to that of the V-cycle whereas with our technique a sharper estimate for the W-cycle can be proved.

As an example of an application of our results, we consider discretizations of anisotropic boundary value problems, that is, problems as

$$
\begin{cases}
-\left(\epsilon\partial_1^2 + \partial_2^2\right)u &= f \quad \text{on } \Omega = (0,1)^2 \\
u &= 0 \quad \text{on } \partial\Omega
\end{cases}
\qquad (\epsilon > 0). \tag{1.1}
$$

Due to the loss of ellipticity, $C_A = C_A(\epsilon)$ can be expected to grow to infinity if $\epsilon \downarrow 0$. Therefore, the "classical" V-cycle bound of Hackbusch and Wittum tends to 1 if $\epsilon \downarrow 0$. However, from numerical experiments we know that it is possible to obtain a MGM that converges uniformly in l and ϵ (robustness) if the smoother is an (almost) direct solver for the equation with $\epsilon = 0$. Under conditions of this kind, robustness of the W-cycle has already been proved (cf. [Wit89b, Ste91]). Up till now, robustness of the V-cycle has not been shown.

In this paper, for the 5-point discretization on all levels of (1.1) and a number of smoothers including (modified) ILU, we show robustness of the V-cycle when the standard 7- or 9-point (no Galerkin approximation) restrictions and prolongations are applied. For the 9-point case, we will give quantitative estimates. Our results concerning robustness are not restricted to this model problem. For example, for semi-positive definite smoothers robustness of the V-cycle is obtained if $\sup_\epsilon \left(C_A(\epsilon) \sup_l \frac{\rho(W_l(\epsilon) - L_l(\epsilon))}{\rho(L_l(\epsilon))} \right) < \infty$.

2   Notations and Basic Assumptions

Using notations as in [Hac85], we assume sequences of linear operators

$$
L_l, S_l : \mathbf{C}^{n_l} \to \mathbf{C}^{n_l}, \; p : \mathbf{C}^{n_l} \to \mathbf{C}^{n_{l+1}} \text{ and } r : \mathbf{C}^{n_{l+1}} \to \mathbf{C}^{n_l} \qquad l \in \mathbf{N}_0 = \{0, 1, 2, \ldots\}
$$

with L_l regular and n_l an increasing function of l. We equip $\mathbf{C}^{n_l}$ with the scaled Euclidian scalar product defined by

$$
(u_l, v_l) = \frac{1}{n_l} \sum_{i=1}^{n_l} (u_l)_i \overline{(v_l)_i}
$$

and norm $\| \cdot \| = (\cdot, \cdot)^{\frac{1}{2}}$. Adjoints relative to $(\cdot, \cdot)$ will be denoted by $*$.

With $K_l = I_l - p L_{l-1}^{-1} r L_l$, $M_l(\nu_1, \nu_2) = S_l^{\nu_2} K_l S_l^{\nu_1}$ is the error amplification operator of the two-grid method (TGM) applied to a given system of linear equations $L_l u_l = f_l$

with ν_1, ν_2 pre- and post-relaxations $u_l^{(i+1)} = S_l u_l^{(i)} + (I_l - S_l)L_l^{-1}f_l$ respectively. With $M_0'(\nu_1, \nu_2, \gamma) := 0$,

$$M_l'(\nu_1, \nu_2, \gamma) = M_l(\nu_1, \nu_2) + S_l^{\nu_2} p M_{l-1}'(\nu_1, \nu_2, \gamma)^\gamma L_{l-1}^{-1} r L_l S_l^{\nu_1}$$

is the error amplification operator of the corresponding MGM with γ recursive calls on the coarse grid.

We will mainly consider the case $\gamma = 1$ (V-cycle) and we will restrict ourselves to $\nu_1 = \nu_2 = \frac{\nu}{2}$. To obtain convergence results for $\gamma = 1$ and $\nu_1 \neq \nu_2$, we refer to [Hac85, theorem 7.2.5]. We will often abbreviate $M_l'(\frac{\nu}{2}, \frac{\nu}{2}, \gamma)$ by M_l' and $M_l(\frac{\nu}{2}, \frac{\nu}{2})$ by M_l.

Basic assumptions that we make throughout this paper are

Assumptions 2.1

(a). $L_l = L_l^* > 0$

(b). $r = p^*$

(c). $L_{l-1} \geq r L_l p \qquad (L_{l-1} = r L_l p$ for Galerkin approximations$)$

(d). $S_l = I_l - W_l^{-1}L_l$, for some regular $W_l = W_l^*$

(e). there exists a $\theta \in [0, 1)$ with $\bigcup_l \sigma(S_l) \subset [-\theta, 1]$ $(\overset{(a),(d)}{\iff} (1+\theta)W_l \geq L_l$ for all $l)$

Above assumptions with (c) replaced by $L_{l-1} = r L_l p$ were also made by Wittum ([Wit89a]) and, with $\theta = 0$, by Hackbusch ([Hac85, §6.4, 7.2]).

Because of (a), on $\mathbf{C}^{n_l}$ we may define the energy scalar product by

$$(u_l, v_l)_E = (L_l u_l, v_l)$$

and norm $\| \cdot \|_E = (\cdot, \cdot)_E^{\frac{1}{2}}$. Adjoints relative to $(\cdot, \cdot)_E$ will be denoted by ^{H}. Because of (a),(b) and (d), we have $K_l^H = K_l$, $S_l^H = S_l$, $M_l^H = M_l$ and so by $p^H = L_{l-1}^{-1} r L_l$, also $M_l'^H = M_l'$. In case $\theta = 0$ in (e), we may formally define $M_l = M_l(\frac{\nu}{2}, \frac{\nu}{2})$ and $M_l' = M_l'(\frac{\nu}{2}, \frac{\nu}{2}, \gamma)$ also for $\nu \in \mathbf{R}_{\geq 0} \backslash 2\mathbf{N}_0$.

In the next section, we will estimate $\rho(M_l') = \|M_l'\|_{E \leftarrow E}$.

3   Estimates of the Contraction Number

Definition 3.1 Throughout the paper, let s be in the same set as ν, that is, $s \in \mathbf{R}_{\geq 0}$ if $\theta = 0$ and $s \in 2\mathbf{N}_0$ if $\theta > 0$. Define

$$c_s = \sup_l \rho(K_l L_l^{-1} W_l S_l^s)$$

and

$$f_s : [-\theta, 1] \times [0, 1] \to [0, \infty] : (\xi, \alpha) \mapsto (\alpha c_s(1 - \xi) + (1 - \alpha)\xi^s)\xi^{\nu - s}.$$

In situations in which the standard *approximation property* (cf. [Hac85])

$$C_A := \sup_l \rho(K_l L_l^{-1})\rho(L_l) < \infty \tag{3.1}$$

holds, it follows from

$$\rho(K_l L_l^{-1} W_l S_l^s) \leq \|K_l L_l^{-1}\| \|W_l S_l^s\| \stackrel{(a)=(e)}{=} \rho(K_l L_l^{-1}) \rho(W_l S_l^s) \tag{3.2}$$

that c_s (and if $s \leq \nu$, thus f_s) is finite under the natural assumption $\sup_l \frac{\rho(W_l S_l^s)}{\rho(L_l)} < \infty$. Another property of the constants c_s is that the function

$$s \mapsto c_s \text{ is non-increasing } \textit{on its domain} \text{ (see definition 3.1)}. \tag{3.3}$$

Indeed, let $s' \geq s \geq 0$ with s, s' even if $\theta > 0$. Then

$$\begin{aligned}
\rho(K_l L_l^{-1} W_l S_l^{s'}) &= \rho(K_l (I_l - S_l)^{-1} S_l^{s'}) \\
&\leq \|S_l^{s/2}(I_l - S_l)^{-1/2} K_l (I_l - S_l)^{-1/2} S_l^{s/2}\|_{E \leftarrow E} \|S_l^{s'-s}\|_{E \leftarrow E} \\
&\leq \rho(K_l L_l^{-1} W_l S_l^s).
\end{aligned}$$

We will use the next proposition concerning the TGM as a lemma for our analysis of the V-cycle but it can clearly also serve to obtain sharp two-grid convergence estimates (cf. [Hac85, theorem 6.4.4] ($\theta = 0$), [Wit89a, criterion 4.1.1] ($\theta \geq 0$)).

Proposition 3.2 *Let $s \leq \nu$ (and as always, $s, \nu \geq 0$ and if $\theta > 0$, s, ν even). Then, with respect to $(\cdot, \cdot)_E$, for all $\alpha \in [0, 1]$ we have*

$$0 \leq M_l \leq f_s(S_l, \alpha).$$

Proof. By assumptions 2.1 (a) and (b), $pL_{l-1}^{-1} r L_l$ and $L_{l-1}^{-1} r L_l p$ are symmetric and semi-positive definite (with respect to $(\cdot, \cdot)_E$). From (c) it follows that $L_{l-1}^{-1} r L_l p \leq I_{l-1}$ and thus by $\rho(p L_{l-1}^{-1} r L_l) = \rho(L_{l-1}^{-1} r L_l p)$ also $p L_{l-1}^{-1} r L_l \leq I_l$. We may conclude that $0 \leq K_l = I_l - p L_{l-1}^{-1} r L_l \leq I_l$ and so

$$0 \leq M_l = S_l^{\nu/2} K_l S_l^{\nu/2} \leq S_l^\nu. \tag{3.4}$$

Furthermore,

$$\begin{aligned}
0 \leq M_l &= S_l^{\frac{\nu-s}{2}}(I_l - S_l)^{\frac{1}{2}}(I_l - S_l)^{-\frac{1}{2}} S_l^{\frac{s}{2}} K_l S_l^{\frac{s}{2}}(I_l - S_l)^{-\frac{1}{2}}(I_l - S_l)^{\frac{1}{2}} S_l^{\frac{\nu-s}{2}} \\
&\leq \rho\left(K_l (I_l - S_l)^{-1} S_l^s\right) S_l^{\nu-s}(I_l - S_l). \tag{3.5}
\end{aligned}$$

Taking a convex combination of (3.4) and (3.5) completes the proof. $\square$

The result of proposition 3.2 with $s = 0$ is very similar to the two-grid estimate of Hackbusch and Wittum. The difference is that for their estimate the parameter c_0 of f_0 should be replaced by $C_A \sup_l \frac{\rho(W_l)}{\rho(L_l)}$ which is larger or equal to c_0 by (3.2).

Using proposition 3.2, we obtain the following recursion for the error amplification operators of the MGM.

Theorem 3.3 *Let $s \leq \nu$. For $\zeta \in [0, 1]$, define*

$$F_s(\zeta) = \min_{\beta \in [0, 1-\zeta^\gamma]} \max_{\xi \in [-\theta, 1]} f_s(\xi, \beta).$$

If, with respect to $(\cdot, \cdot)_E$, for some $\zeta_{l-1} \in [0, 1]$,

$$0 \leq M'_{l-1} \leq \zeta_{l-1} I_{l-1},$$

then

$$0 \leq M_l' \leq F_s(\zeta_{l-1})I_l. \tag{3.6}$$

In particular, using $M_0' = 0$, we have

$$\sup_l \|M_l'\|_{E \leftarrow E} \leq \zeta^* := \min \{\zeta \in [0,1] : F_s(\zeta) = \zeta\}.$$

(Note that ζ^* is well defined because $F_s(1) = 1$).

Proof. It holds that $M_l' = M_l + S_l^{\nu/2} p M_{l-1}'^{\gamma} L_{l-1}^{-1} r L_l S_l^{\nu/2}$. From $p^H = L_{l-1}^{-1} r L_l$, $S_l^H = S_l$, $M_{l-1}' \geq 0$ and $M_l \geq 0$ by proposition 3.2, it follows that $M_l' \geq 0$. Using $M_{l-1}' \leq \zeta_{l-1} I_{l-1}$ and $M_l \leq f_s(S_l, \alpha)$ for $\alpha \in [0,1]$ by proposition 3.2, we have

$$\begin{aligned}
M_l' &\leq M_l + \zeta_{l-1}^{\gamma} S_l^{\nu/2} p L_{l-1}^{-1} r L_l S_l^{\nu/2} = (1 - \zeta_{l-1}^{\gamma}) M_l + \zeta_{l-1}^{\gamma} S_l^{\nu} \\
&\leq (1 - \zeta_{l-1}^{\gamma}) f_s(S_l, \alpha) + \zeta_{l-1}^{\gamma} S_l^{\nu} = f_s(S_l, (1 - \zeta_{l-1}^{\gamma})\alpha).
\end{aligned}$$

Since $\alpha \in [0,1]$ is arbitrary and $-\theta I_l \leq S_l \leq I_l$, the inequality $M_l' \leq F_s(\zeta_{l-1})I_l$ follows. $\square$

The proof of the following corollary concerning the V-cycle will be given in [Ste92].

Corollary 3.4 *Let $\gamma = 1$ and $\nu \geq 1$. Then for $s \leq \frac{\nu+1}{2}$ or $s = \nu$, $\zeta^* = \dfrac{c_s}{c_s + \min\left\{\nu, \frac{1-\theta^\nu}{(1+\theta)\theta^{\nu-s}}\right\}}$*

and so

$$\sup_l \|M_l'\|_{E \leftarrow E} \leq \dfrac{c_s}{c_s + \min\left\{\nu, \frac{1-\theta^\nu}{(1+\theta)\theta^{\nu-s}}\right\}}.$$

Remarks 3.5 (a). For $s \leq \frac{\nu+1}{2}$ and ν sufficiently large or θ sufficiently small, in particular for $\theta = 0$, it holds that $\min\left\{\nu, \frac{1-\theta^\nu}{(1+\theta)\theta^{\nu-s}}\right\} = \nu$. So, at least in this case, the result of corollary 3.4 for positive s is never worse than for $s = 0$ since $s \mapsto c_s$ is non-increasing (cf. (3.3)). Note that we can not expect that $\lim_{s \to \infty} c_s = 0$ because otherwise we would have proved that $\sup_l \|M_l'\|_{E \leftarrow E} = o(\nu^{-1})$ (take $s = \frac{\nu+1}{2}$ or, if $\theta > 0$, $s = 2\lfloor \frac{\nu+1}{4} \rfloor \in 2\mathbf{N}$) and by using proposition 3.2 also that $\sup_l \|M_l\|_{E \leftarrow E} = o(\nu^{-1})$. For a number of model problems it can be shown that $\sup_l \|M_l\|_{E \leftarrow E} \sim \nu^{-1}$ (cf. e.g. [Hac85, theorem 2.4.4 (and exercise 2.7.3)]).

For $s = \nu \geq 1$, it holds that $\min\left\{\nu, \frac{1-\theta^\nu}{(1+\theta)\theta^{\nu-s}}\right\} = \frac{1-\theta^\nu}{1+\theta}$.

(b). For each s, the function $\gamma \mapsto \zeta^* = \zeta^*(\gamma)$ is decreasing. In [Ste92], for $s \in \{0,1,2\}$ we will determine $\zeta^*(2)$ (W-cycle) and $\zeta^*(\infty)$ (TGM).

(c). In order to compute an upper bound for $\|M_l'\|_{E \leftarrow E}$ for fixed l, the definition $c_s = \sup_l \rho(K_l L_l^{-1} W_l S_l^s)$ may be replaced by $c_s = \sup_{k \leq l} \rho(K_l L_l^{-1} W_l S_l^s)$.

(d). Another way to get a sharper bound for fixed l which also can be used for a general l-grid method on "level" $l + q$ for arbitrary $q \in \mathbf{N}_0$, is to use the recursion of theorem 3.3 l times instead of considering the fixed point ζ^* (cf. [Hac85, table 7.2.1]). If q is given, then on each level $q + 1 \leq k \leq l + q$ one may use the definition $c_s = \rho(K_k L_k^{-1} W_k S_k^s)$.

In cases where the approximation property (3.1) holds, we have $c_s \leq C_A \sup_l \frac{\rho(W_l S_l^s)}{\rho(L_l)}$ (cf. (3.2)). In combination with corollary 3.4 this yields the following estimates for the contraction number

$$\sup_l \|M_l'\|_{E \leftarrow E} \leq \frac{C_A \sup_l \frac{\rho(W_l S_l^s)}{\rho(L_l)}}{C_A \sup_l \frac{\rho(W_l S_l^s)}{\rho(L_l)} + \min\left\{\nu, \frac{1-\theta^\nu}{(1+\theta)\theta^{\nu-s}}\right\}}. \tag{3.7}$$

For $s = 0$, (3.7) equals the estimate that was found by Hackbusch ([Hac85, theorem 7.2.2]) ($\theta = 0$) and Wittum ([Wit89a, criterion 4.2.1]) ($\theta \geq 0$). For $s = 1$, and thus $\theta = 0$, (3.7) can be deduced from results of Mandel, McCormick and Bank ([MMB87, theorems 4.4 and 5.2, remark 4.4]) *if $L_{l-1} = rL_l p$*. The results for $s \notin \{0,1\}$ are new.

If $\theta = 0$, it follows from $W_l > W_l - L_l \geq 0$ and thus $\rho(W_l S_l) = \rho(W_l - L_l) < \rho(W_l)$ that (3.7) with $s = 1$ always yields a sharper upper bound than (3.7) with $s = 0$. A quantitative comparison of both bounds will be made in example 4.2 (tables 4.2 and 4.3). Note that in contrast to $s = 0$, for $s = 1$ (and $s > 1$), the value of $\sup_l \frac{\rho(W_l S_l^s)}{\rho(L_l)}$ is related to the quality of the smoother.

For discretizations of anisotropic boundary value problems, that is, problems as

$$\begin{cases} -(\epsilon \partial_1^2 + \partial_2^2)\, u &= f \quad \text{on } \Omega = (0,1)^2 \\ u &= 0 \quad \text{on } \partial\Omega \end{cases} \quad (\epsilon > 0), \tag{3.8}$$

the constant $C_A = C_A(\epsilon)$ can be expected to grow to infinity if ϵ tends to zero. When a suitable smoother is applied the growth of $C_A(\epsilon)$ is compensated by a decrease of $\sup_l \frac{\rho(W_l(\epsilon) - L_l(\epsilon))}{\rho(L_l(\epsilon))}$ so that the product of these two factors is bounded uniformly in ϵ. If in addition the smoother is semi-positive definite, then the application of (3.7) with $s = 1$ shows that the contraction number of the V-cycle is bounded below 1 uniformly in l *and* ϵ (robustness).

Note that for any smoother that satisfies assumptions 2.1 (d) and (e), it holds that $2W_l > L_l$ and so $\sup_l \frac{\rho(W_l)}{\rho(L_l)} > \frac{1}{2}$. Therefore, robustness can not be shown with (3.7) with $s = 0$.

To show robustness for symmetric smoothers with also negative eigenvalues, as the ILU smoothers, (3.7) with $s = 2$ can be applied (since corollary 3.4 is also valid for $s = \nu$, this is also possible for $\nu = 2$). A necessary condition is that assumptions 2.1 (d) is satisfied with $\sup_\epsilon \theta(\epsilon) < 1$ since otherwise $\limsup_{\epsilon \downarrow 0} \min\left\{\nu, \frac{1-\theta(\epsilon)^\nu}{(1+\theta(\epsilon))\theta(\epsilon)^{\nu-s}}\right\} = 0$. By

$$\rho(W_l S_l^2) = \rho((W_l - L_l)S_l) \leq \rho(W_l - L_l)\|S_l\|, \tag{3.9}$$

robustness is now shown if $\sup_\epsilon \left(C_A(\epsilon) \sup_l \frac{\rho(W_l(\epsilon) - L_l(\epsilon))}{\rho(L_l(\epsilon))}\right) < \infty$ and $\sup_{\epsilon,l} \|S_l(\epsilon)\| < \infty$. So, in comparison with $\theta = 0$ (and $s = 1$), an additional condition has to be checked. (Note that although also for $\theta > 0$ we could define $c_1 = \sup_l \rho(K_l L_l^{-1} W_l S_l)$, in that case we do not have $c_2 \leq c_1$ ($\leq \sup_\epsilon \left(C_A(\epsilon) \sup_l \frac{\rho(W_l(\epsilon) - L_l(\epsilon))}{\rho(L_l(\epsilon))}\right)$ (cf. (3.3)).)

4 Applications to the model anisotropic equation (3.8)

In this section, we consider the standard 5-point discretization on all levels of the model anisotropic equation (3.8). As examples of applications of our V-cycle bounds, for standard 7- *or* 9-point (no Galerkin approximation) restrictions and prolongations and a number of smoothers we show robustness of the V-cycle MGM.

Example 4.1 Let $L_l(\epsilon) \triangleq \frac{1}{h_l^2} \begin{bmatrix} & -1 & \\ -\epsilon & 2+2\epsilon & -\epsilon \\ & -1 & \end{bmatrix} : \mathbf{C}^{n_l} \to \mathbf{C}^{n_l}$ where $h_l = 2^{-(l+1)}$ and

$n_l = (h_l^{-1} - 1)^2$, that is, $L_l(\epsilon)$ is the standard 5-point discretization of (3.8) on the grid

$(0,1)^2 \cap h_l \mathbf{Z}^2$; $p \triangleq \frac{1}{2} \begin{bmatrix} 1 & 1 \\ 1 & 2 & 1 \\ 1 & 1 \end{bmatrix} : \mathbf{C}^{n_l} \to \mathbf{C}^{n_{l+1}}$ and $r \triangleq \frac{1}{8} \begin{bmatrix} 1 & 1 \\ 1 & 2 & 1 \\ 1 & 1 \end{bmatrix} : \mathbf{C}^{n_{l+1}} \to \mathbf{C}^{n_l}$.

In this situation we have $L_l(\epsilon) = L_l(\epsilon)^*$ (assumptions 2.1 (a)), $r = p^*$ ((b)) and $L_{l-1}(\epsilon) = rL_l(\epsilon)p$ ((c)). In [Ste91, proposition 2.1 and remark 2.2], it is proved that

$$\rho(K_l(\epsilon)L_l(\epsilon)^{-1})\rho(L_l(\epsilon)) = \mathcal{O}\left(\min\{\epsilon^{-1}, h_l^{-2}\}\right) \quad (\epsilon, h_l \downarrow 0) \tag{4.1}$$

and thus $C_A(\epsilon) = \mathcal{O}(\epsilon^{-1})$.

The symmetric lexicographical y-(zebra)-line Gauss-Seidel smoother satisfies (d) and (e) with $\theta = 0$. It is easily checked that $\sup_l \frac{\rho(W_l(\epsilon) - L_l(\epsilon))}{\rho(L_l(\epsilon))} = \mathcal{O}(\epsilon)$ ($\epsilon \downarrow 0$). Estimate (3.7) with $s = 1$ now shows robustness of the V-cycle. From (4.1), it appears that for fixed l the upper bound for the contraction number even tends to zero when $\epsilon \downarrow 0$ (cf. remarks 3.5 (c)).

The 5-point ILU-smoother satisfies (d) and (e). However, the smallest eigenvalue of S_l tends to -1 when $\epsilon \downarrow 0$ and therefore no robustness of the V-cycle can be shown using corollary 3.4. From numerical experiments (cf. [Wit89a, Wit89b]), it can be observed that this smoother does not yield a robust method either.

In order to obtain a robust method, Wittum ([Wit89b]) proposed a modification of the ILU-smoother controlled by some parameter β which he called ILU$_\beta$. As shown in [Wit89b], for $\beta > 0$ this smoother satisfies (d), (e) with $\theta(\epsilon) < \frac{1-\beta}{1+\beta} < 1$, $\rho(W_l(\epsilon) - L_l(\epsilon)) \leq 2(1+\beta)\epsilon h_l^{-2} = \mathcal{O}(\epsilon)\rho(L_l(\epsilon))$ and $\sup_{\epsilon,l} \|S_l(\epsilon)\| \leq 1$. Robustness of the V-cycle is now shown by estimate (3.7) with $s = 2$ in combination with (3.9) or, if $\beta \geq 1$, by (3.7) with $s = 1$. Again, by (4.1), the upper bound for the contraction number for fixed l even tends to zero if $\epsilon \downarrow 0$.

Example 4.2 Let $L_l(\epsilon)$ be as in example 4.1 but now let $p \triangleq \frac{1}{4} \begin{bmatrix} 1 & 2 & 1 \\ 2 & 4 & 2 \\ 1 & 2 & 1 \end{bmatrix}$ and

$r \ (= p^*) \triangleq \frac{1}{16} \begin{bmatrix} 1 & 2 & 1 \\ 2 & 4 & 2 \\ 1 & 2 & 1 \end{bmatrix}$. For example by using the bases of eigenfunctions of $L_l(\epsilon)$

and $L_{l-1}(\epsilon)$ (cf. [ST82, lemma 7.1], [MMB87, §5.8.4]), one may check

$$L_{l-1}(\epsilon) - rL_l(\epsilon)p \triangleq \frac{1}{4}h_{l-1}^{-2}(1+\epsilon)\begin{bmatrix} -1 & 2 & -1 \end{bmatrix}\begin{bmatrix} -1 \\ 2 \\ -1 \end{bmatrix} > 0 \ .$$

165

So $L_{l-1}(\epsilon)$ is not a Galerkin approximation but it does satisfy our assumptions 2.1 (c). Again by using the bases of eigenfunctions, it can be computed that

$$h_l^{-2}\rho(K_l(\epsilon)L_l(\epsilon)^{-1}) = \mathcal{O}\left(\min\{\epsilon^{-1}, h_l^{-2}\}\right) \quad (\epsilon, h_l \downarrow 0). \tag{4.2}$$

We may conclude that the symmetric lexicographical y-(zebra)-line Gauss-Seidel smoother and, for $\beta > 0$, the ILU$_\beta$ smoother yield robust V-cycles and moreover, that for fixed l the contraction numbers tend to zero when $\epsilon \downarrow 0$.

For this example we are able to compute $h_l^{-2}\rho(K_l(\epsilon)L_l(\epsilon)^{-1})$ numerically. In table 4.1 some values are given.

Table 4.1: $h_l^{-2}\rho(K_l(\epsilon)L_l(\epsilon)^{-1})$

$h_l \setminus \epsilon$	1	.5	.1	10^{-2}	10^{-4}	10^{-8}
$\frac{1}{16}$	.494	.963	4.19	17.1	25.9	26.0
$\frac{1}{32}$	.498	.990	4.77	33.7	102	104
$\frac{1}{64}$	.500	.998	4.94	44.6	383	415
$\frac{1}{128}$	.500	.999	4.99	48.5	1250	1660
$\frac{1}{256}$	.500	1.00	5.00	49.6	2850	6640

For the symmetric lexicographical y-line Gauss-Seidel smoother, it holds that

$$h_l^2(W_l(\epsilon) - L_l(\epsilon)) \triangleq \begin{bmatrix} & \\ & \epsilon \end{bmatrix} \begin{bmatrix} -1 & 2 + 2\epsilon & -1 \end{bmatrix}^{-1} \begin{bmatrix} & \epsilon \\ & \end{bmatrix}$$

and so

$$h_l^2\rho(W_l(\epsilon) - L_l(\epsilon)) \le \frac{\epsilon^2}{2\epsilon + 4\sin^2(h_l\pi/2)}. \tag{4.3}$$

By using corollary 3.4 with $s = 1$, remarks 3.5 (c), (3.2), (4.3) and the numbers from table 4.1, we have computed the upper bounds for $\|M_l'(1,1,1)(\epsilon)\|_{E \leftarrow E}$ ($\nu = 2$, $\gamma = 1$) given in table 4.2.

Table 4.2: Upper bounds for $\|M_l'(1,1,1)(\epsilon)\|_{E \leftarrow E}$ obtained with $s = 1$.

$h_l \setminus \epsilon$	1	.5	.1	10^{-2}	10^{-4}	10^{-8}
$\frac{1}{16}$	.108	.104	.0808	.0144	3.35_{10-6}	3.39_{10-14}
$\frac{1}{32}$	.110	.109	.102	.0539	5.17_{10-5}	5.39_{10-13}
$\frac{1}{64}$	.111	.111	.109	.0906	7.34_{10-4}	8.62_{10-12}
$\frac{1}{128}$	.111	.111	.111	.105	7.71_{10-3}	1.38_{10-10}
$\frac{1}{256}$	.111	.111	.111	.110	0.0391	2.20_{10-9}

Using $\rho(W_l(\epsilon)) \ge \rho(L_l(\epsilon)) = (4 + 4\epsilon)\sin^2((1 - h_l)\pi/2)$, we have also computed lower bounds for the "classical" upper bounds of $\|M_l'(1,1,1)(\epsilon)\|_{E \leftarrow E}$ that are obtained by corollary 3.4 with $s = 0$, remarks 3.5 (c), (3.2) and table 4.1. The values are given in table 4.3.

166

Table 4.3: Upper bounds for $\|M_l'(1,1,1)(\epsilon)\|_{E\leftarrow E}$ obtained with $s = 0$.

$h_l \setminus \epsilon$	1	.5	.1	10^{-2}	10^{-4}	10^{-8}
$\frac{1}{16}$	$\geq .662$	$\geq .741$	$\geq .901$	$\geq .972$	$\geq .981$	$\geq .981$
$\frac{1}{32}$	$\geq .665$	$\geq .748$	$\geq .913$	$\geq .986$	$\geq .995$	$\geq .995$
$\frac{1}{64}$	$\geq .666$	$\geq .749$	$\geq .916$	$\geq .989$	$\geq .999$	$\geq .999$
$\frac{1}{128}$	$\geq .667$	$\geq .750$	$\geq .916$	$\geq .990$	≥ 1.00	≥ 1.00
$\frac{1}{256}$	$\geq .667$	$\geq .750$	$\geq .917$	$\geq .990$	≥ 1.00	≥ 1.00

From the tables 4.2 and 4.3 we see that the bounds obtained with the new estimate with $s = 1$ are at least a factor 6 less than the bounds obtained with the already existing estimate with $s = 0$. Moreover, the results clearly illustrate that with the estimate with $s = 1$ robustness can be shown whereas with the estimate with $s = 0$ this is not possible.

Note that by (4.2) and (4.3), for *fixed l* the bounds obtained with $s = 1$ are of order ϵ. Because of $\|S_l\| \leq \rho(W_l^{-1})\rho(W_l - L_l) \leq \rho(L_l^{-1})\rho(W_l - L_l)$ and, for *fixed l*, $\rho(L_l(\epsilon)^{-1}) = \left((4 + 4\epsilon)\sin^2(h_l\pi/2)\right)^{-1} = \mathcal{O}(1)$, we can find asymptotically better bounds of order ϵ^3, or in general $\epsilon^{2\nu-1}$, by taking $s = \nu$ in corollary 3.4. By $M_l' \leq S_l^\nu$ which follows from the proof of theorem 3.3 (take $\alpha = 0$ and use $M_{l-1}' \leq I_{l-1}$), for *fixed l* even $\|M_l'(1,1,1)(\epsilon)\|_{E\leftarrow E} = \mathcal{O}(\epsilon^{2\nu})$ is valid.

References

[Hac85]   W. Hackbusch. *Multi-Grid Methods and Applications*. Springer-Verlag, Berlin, 1985.

[Hac91]   W. Hackbusch. *Iterative Lösung großer schwachbesetzter Gleichungssysteme*. B.G. Teubner, Stuttgart, 1991.

[MMB87]   J. Mandel, S. McCormick, and R. Bank. Variational multigrid theory. In S. McCormick, editor, *Multigrid Methods*, chapter 5. SIAM, Philadelphia, 1987.

[ST82]   K. Stüben and U. Trottenberg. Multigrid methods: Fundamental algorithms, model problem analysis and applications. In Hackbusch W. and U. Trottenberg, editors, *Multigrid Methods*, Proceedings, Köln-Porz 1981, pages 1–176, Berlin, 1982. Lecture Notes in Mathematics 960, Springer-Verlag.

[Ste91]   R.P. Stevenson. On the robustness of multi-grid applied to anisotropic equations: Smoothing- and approximation-properties. Preprint 685, University of Utrecht, September 1991. Submitted to Numerische Mathematik.

[Ste92]   R.P. Stevenson. Sharp estimates of the multi-grid contraction number including the V-cycle. In preparation, 1992.

[Wit89a]   G. Wittum. Linear iterations as smoothers in multigrid methods: Theory with applications to incomplete decompositions. *IMPACT Comput. Sci. Eng.*, 1:180–215, 1989.

[Wit89b]   G. Wittum. On the robustness of ILU smoothing. *SIAM J. Sci. Stat. Comput.*, 10(4):699–717, July 1989.

On ordering strategies in a multigrid algorithm

Stefan Turek
Institut für Angewandte Mathematik
Universität Heidelberg
Im Neuenheimer Feld 294
D-6900 Heidelberg

Introduction

We present some numerical results for the influence of grid renumbering strategies on multigrid smoothers applied to the Stokes equations. Using discrete divergence-free finite elements we get positive definite stiffness matrices, which allow the use of standard smoothers such as Jacobi-, Gauß–Seidel- and ILU–relaxation. For different ordering strategies (finite element two level ordering, geometrical ordering (row- and linewise), Cuthill–McKee–algorithm) we show the costs of the renumbering and the convergence rates for solving the Stokes equations on several types of domains and subdivisions. These convergence results carry over to an algorithm for solving the stationary and instationary Navier–Stokes equations.

Let Ω be a bounded domain in $\mathbf{R}^2$ with boundary $\partial\Omega$. We consider the usual linear Stokes problem

Find a pair $\{\mathbf{u}, p\}$, such that

$$-\Delta\mathbf{u} + \nabla p = \mathbf{f} \quad , \quad \nabla\cdot\mathbf{u} = 0 \quad \text{in } \Omega, \quad \mathbf{u} = \mathbf{g} \quad \text{on } \partial\Omega, \tag{St}$$

where $\{\mathbf{u}, p\}$ are the velocity and pressure, respectively, of a viscous incompressible fluid contained in Ω (cf. [3]). Using quadrilateral *rotated bilinear* discrete divergence–free finite elements for the discretisation (see [7], and also [4]) the corresponding stiffness matrices are positive definite with a condition number behaving like $O(h^{-4})$. The pressure is eliminated, and the resulting set of unknowns consists of approximations for the streamfunction values in the vertices and for the tangential velocities in the midpoints of the underlying triangulation (cf. [7]).

All our computations are based on **FEAT**-routines (Finite Element Analysis Toolbox in 2 and 3 dimensions) (see [1]) and use the *standard compact storage technique* for all matrices (a long vector for the matrix elements with two index pointers) and the *two level ordering* for the subdivisions. This means that starting from a macro decomposition, the refined grids are obtained by connecting the opposite midpoints. The numbers of the *old* gridpoints are preserved, while the *new* vertices on the finer grid get the numbers of the midpoints on the coarser grid (compare the following example).

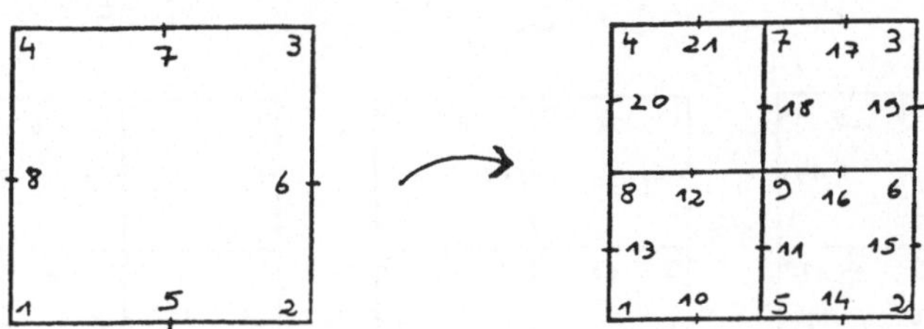

Figure 1: Example for the used grid numbering

This data structure is well suited for the elementwise calculation of the matrices corresponding to the bilinear forms and for the elementwise realization of the grid transfer routines in a multigrid algorithm. Our multilevel method (cf. [7],[9]) consists of the standard components *smoothing* (Jacobi-, Gauß–Seidel- and ILU-relaxation), *coarse grid correction* (macroelementwise divergence–free interpolation) and *adaptive step length control* of the correction. Since only the smoothers, the exact solver and the defect computation are influenced by the used numbering of the unknowns, we first compute two index vectors for the renumbering and reorder only the stiffness matrix. Then, in our multigrid procedure before smoothing, exact solution and defect computation, we apply these index vectors for reordering the solution and auxiliary vectors and renumber back after performing the matrix–vector applications. The original data structure is preserved and the renumbering has no influence onto the intergrid operators. In our tests we used the following 6 ordering strategies (see figure 2):

1. Two Level Ordering (**TLO**) (this means no reordering)

2. Sorting of the x–coordinate (**GR**) (this means "rowwise" numbering, first the vertices then the midpoints)

3. Sorting of the y–coordinate (**GL**) (analogous "linewise" numbering)

4. Modified sorting of the x–coordinate (**MGR**) ("rowwise" numbering, but vertices and midpoints together)

5. Modified sorting of the y–coordinate (**MGL**) (analogous "linewise" numbering)

6. Cuthill–McKee–algorithm (**CM**) (see [2],[6]) (vertices and midpoints together)

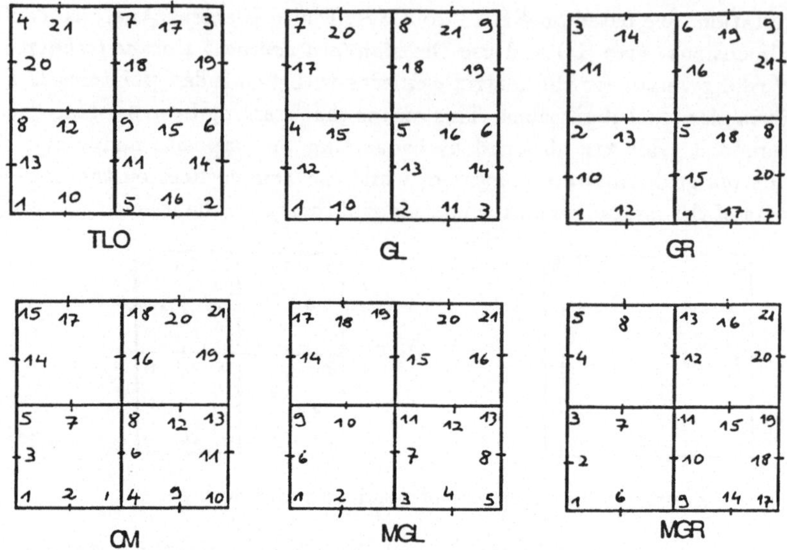

Figure 2: Examples for the renumbering strategies

We can classify these strategies in two groups: The *geometrical* sorting (2,3,4,5) and the *local structure* sorting (6). For the *geometrical* sorting we have to compare the size of the coordinates of all vertices and/or the midpoints. This costs (if N is the number of unknowns) between $O(N \log N)$ arithmetic operations (if we have really a lot of main memory) and, in the worst case, $O(N^2)$ operations. In contradiction, the *local structure* sorting evaluates the matrix pattern and creates the index pointers in $O(N)$ operations by a recursive marching process. The advantages of this method are the much less computational costs and memory demands in comparison with the usual *geometrical* sorting, but what is the improvement for the linear algebra? For line- or rowwise numbering on rectangular grids we know already the strong improvements, at least for scalar Poisson- or Convection–Diffusion problems. Therefore, we will show for some test configurations the needed computational time for evaluating the index vectors on a SUN 4/260 workstation and the corresponding convergence rates for solving the Stokes equations.

As test examples we consider the standard "Driven Cavity Stokes" problem (cf. [7],[9]) on the unit square with the triangulations DC1 and DC2 (see figure 3, which shows the used grids for level 1 and level 3), and the (nonphysical) "flow in a car" problem (with parabolic inflow in front of the "car" and parabolic outflow at the exhaust), using the both triangulations CAR1 and CAR2 (see figure 3).

Figure 3: Grids for "Driven Cavity" DC1 and DC2 and "flow in car" CAR1 and CAR2

The following figure 4 shows the *normalized* evaluation times (divided by $4^{(\text{levelnumber}-1)}$) for computing the index vectors on the levels 1 to 5. In the figures 5 – 8 we present the convergence rates of our multigrid method for Jacobi- (jac), Gauß–Seidel- (gs) and ILU–relaxation (ilu), with 1, 2 or 4 pre- and postsmoothing steps (from top to bottom), in each case on the levels 2 to 5 (from left to right).

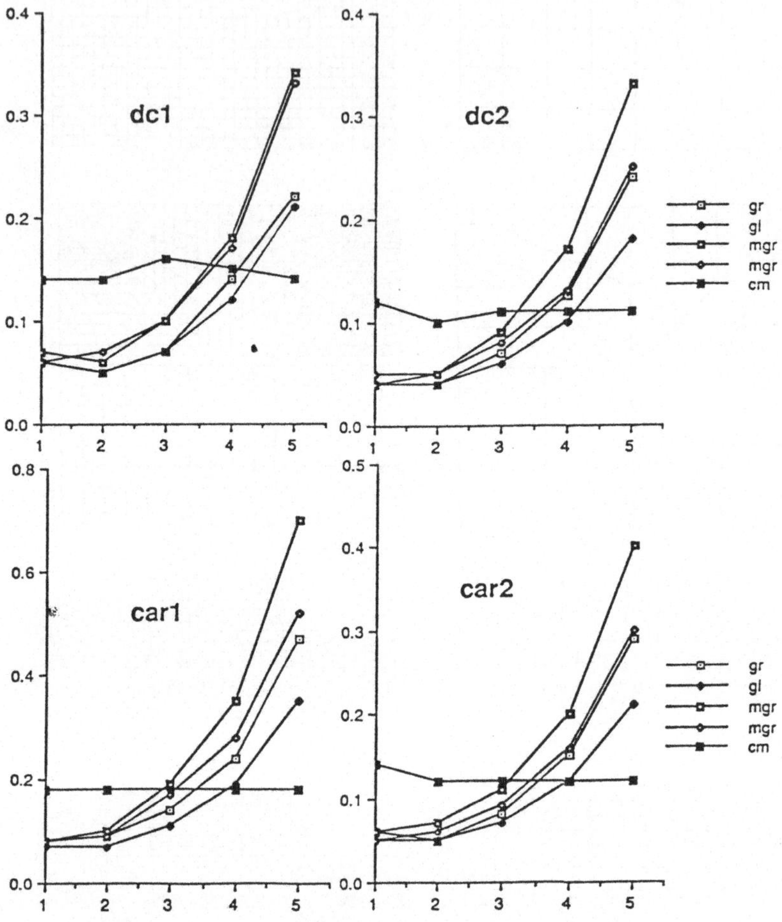

Figure 4: *Normalized* computational times in seconds for the index vectors (level 1 - 5)

Figure 5 shows the convergence rates for the cartesian mesh DC1 without renumbering. The proposed reordering strategies lead to nearly the same results, so we cancel their visualization. The ILU–relaxation yields the best rates, but Gauß–Seidel and Jacobi are of the same order, and especially the Gauß–Seidel–relaxation is even superior when using only 1 smoothing step.

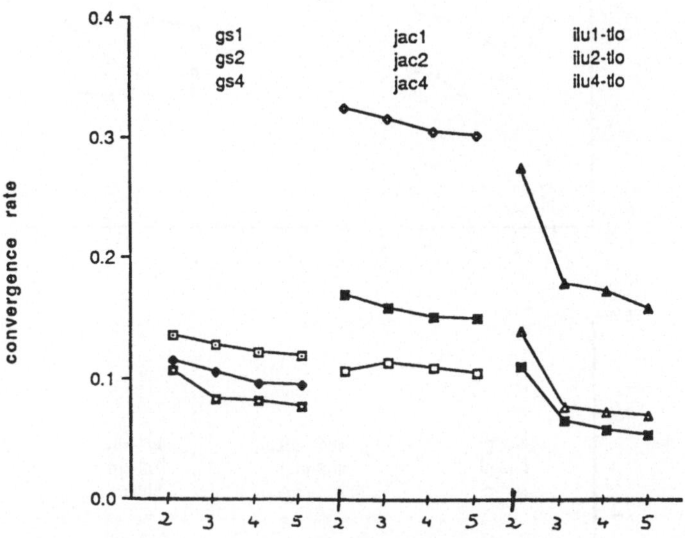

Figure 5: Convergence rates for grid DC1

In figure 6 the corresponding results for the "locally refined" mesh DC2 are presented. Since the x–coordinate and y–coordinate sorting lead to nearly the same results, we show only the rates for the algorithms GY, MGY and CM. The reordering has only influence on the ILU–method, especially if just 1 smoothing step is used. But again, the Gauß–Seidel–relaxation seems to be favourable, since we get very good convergence rates without renumbering. The Jacobi–smoother, too, brings good results. It is remarkable, that the Gauß–Seidel–method even leads to somewhat worse convergence rates if reordering is used.

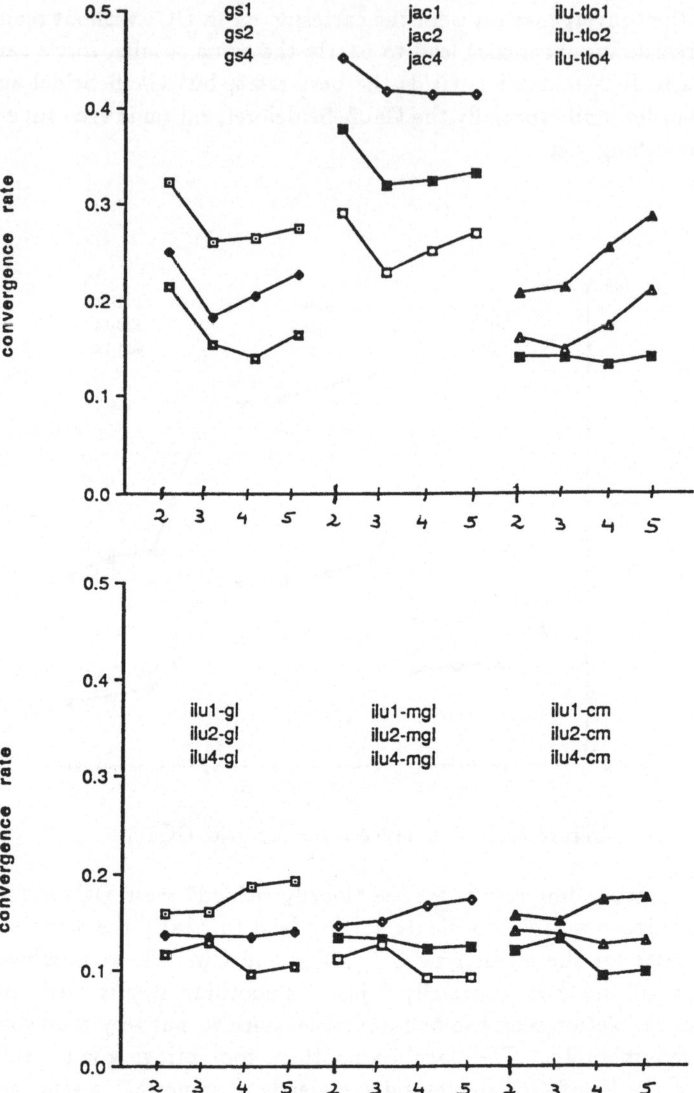

Figure 6: Convergence rates for grid DC2

Now we leave the unit square and show the results for triangulation CAR1 (see figure 7), which is still rather regular. Again, the results for Gauß–Seidel and Jacobi are very good, both without renumbering, while for the ILU–relaxation the reordering should be used. Here we see again, that the Cuthill–McKee–algorithm yields nearly the same results as the "classical" geometrical sorting. The improvement for 1 and 2 smoothing steps is remarkable.

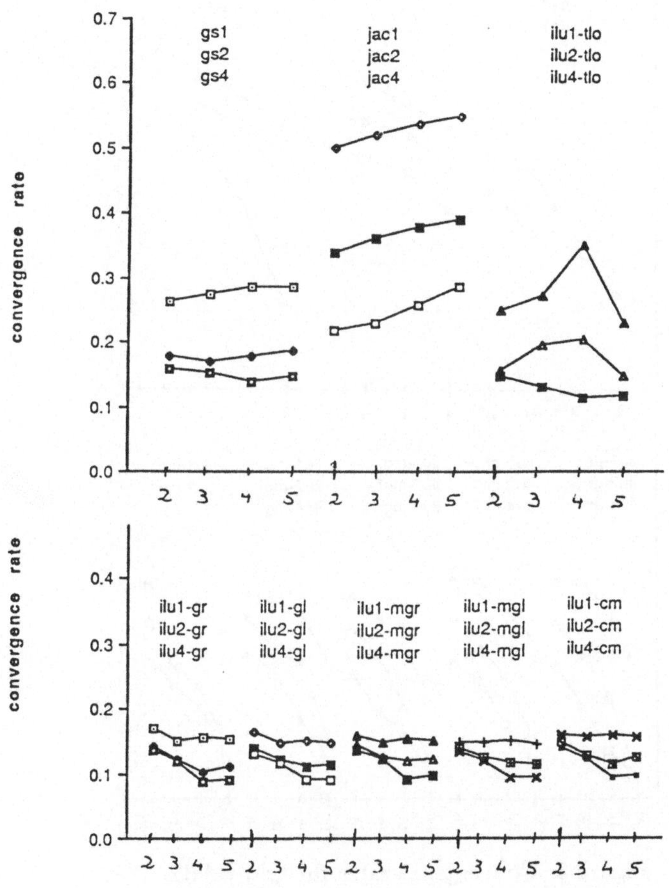

Figure 7: Convergence rates for grid CAR1

Finally we present the rates for the grid CAR2, where the local gridsize is strongly
varying. Here we can see, that the simple methods like Gauß–Seidel and Jacobi have
some problems, only the ILU–method with renumbering shows better results. Again
we see no differences for x- and y–sorting and, fortunately, also the cheaper Cuthill–
McKee–algorithm leads to nearly the same rates.

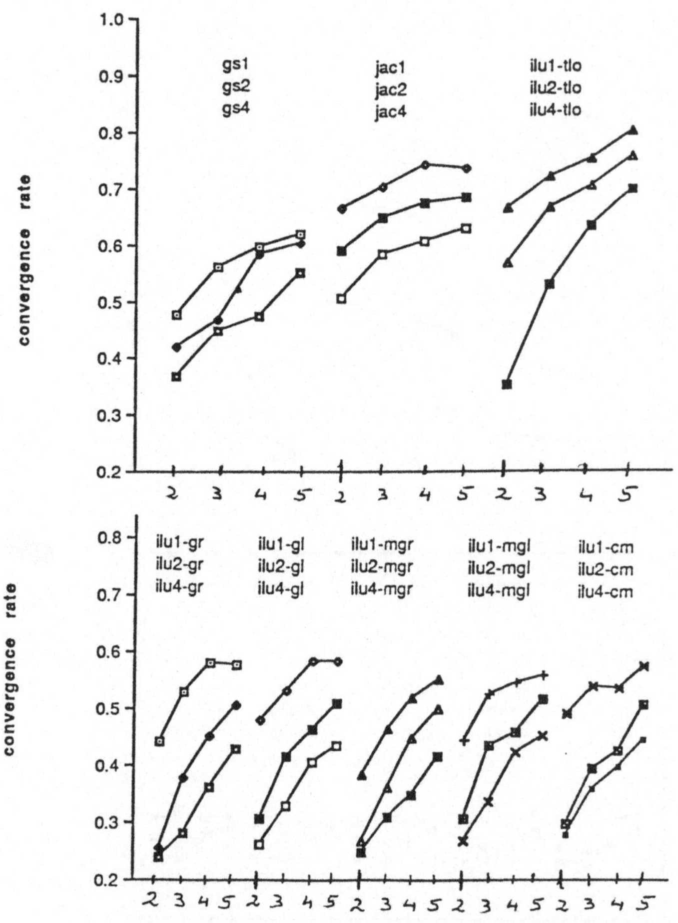

Figure 8: Convergence rates for grid CAR2

From these results we can state:

1. The Jacobi–relaxation is, of course, independent of the used numbering and leads to useful results, if the underlying mesh is not too irregular.

2. The Gauß–Seidel–method should be used without any reordering, since the results often are getting somewhat worse. If not too irregular grids are used, this method seems to be favourable (comparing arithmetic operations and storage requirements), only on the very irregular mesh CAR2 the results are not satisfactory.

3. For the ILU–relaxation certain renumbering strategies should be used. The proposed Cuthill–McKee–algorithm is preferable, since the costs for evaluating the index vectors are minimal in comparison to the other methods. Using geometrical sorting, the "separated" sorting (first the vertices, then the midpoints) has to be preferred, because of the minimal costs.

Nearly the same results can be achieved if these renumbering strategies are used for scalar Poisson equations with nonconforming finite elements (cf. [8]) and for the biharmonic equation, discretized with the nonconforming Morley space (cf. [5]). Also in 3-D, where the *local structure* sorting is much cheaper, the Cuthill–McKee–algorithm is preferable. Our results are also relevant for the incompressible Navier–Stokes equations (cf. [3])

$$A(\mathbf{u})\mathbf{u} := -\nu\Delta\mathbf{u} + \mathbf{u}\nabla\mathbf{u} + \nabla p = \mathbf{f}, \quad \nabla\cdot\mathbf{u} = 0, \quad \text{in } \Omega, \quad \mathbf{u} = \mathbf{g}, \quad \text{on } \partial\Omega,$$

where we use (see [7]) a fixpoint–defect correction–method in the form

$$\mathbf{u}^{n+1} = \mathbf{u}^n + \omega[\tilde{A}(\mathbf{u}^n)]^{-1}(\mathbf{f} - A(\mathbf{u}^n)\mathbf{u}^n).$$

Here, $\tilde{A}(.)$ corresponds to an upstream discretization for the convective terms and is a M-Matrix, so we have to invert a linear Oseen matrix in each iteration step, what can be performed very efficiently by our multigrid method. In all test cases we reach for these unsymmetric problems nearly the same convergence results as for the Stokes problem, when the Gauß–Seidel–relaxation is used as smoother. Further, in connection with an appropriate time stepping scheme (implicit Euler, Crank-Nicolson, Fractional Step scheme) also the instationary Navier–Stokes equations can be solved with nearly the same linear convergence rates.

Conclusion

If the ILU–method is used as a smoother in a multigrid algorithm some appropriate grid renumbering strategies should be used to improve the convergence rates. Since geometrical sorting has a large requirement of storage size and arithmetic costs ($O(N \log N)$ – $O(N^2)$), an alternative method is the Cuthill–McKee–algorithm, which was originally proposed for bandwidth optimization. This method directly uses the finite element space structure (scalar nonconforming, discrete divergence–free, Morley space), and its runtime analysis and the numerical results show an $O(N)$ operations account. Since in 3-D the geometrical sorting is much more expensive, this CM–algorithm seems to be favourable if ILU–decompositions are used. But considering all performed tests (also for the Navier–Stokes equations, on not too irregular meshes), the Gauß–Seidel–relaxation is my preferred smoother because of its small costs (no additional matrix and no renumbering).

References

[1] Blum, H., Harig, J., Müller S.: **FEAT** . *Finite element analysis tools. Release 1.1. User Manual*, Technischer Report Nr. 554, SFB 123, Universität Heidelberg, 1990

[2] Cuthill, E., McKee, J.: *Reducing the bandwidth of sparse symmetric matrices*, in Proc. ACM Nat. Conf., 157–172, New York 1969

[3] Girault, V., Raviart, P.A.: *Finite Element Methods for Navier–Stokes equations*, Springer, Berlin–Heidelberg 1986

[4] Rannacher, R., Turek, S.: *A simple nonconforming quadrilateral Stokes element*, Technischer Report Nr. 567, SFB 123, Universität Heidelberg, April 1990

[5] Schreiber, P.: *Ein robustes und effizientes Mehrgitterverfahren zur Lösung der biharmonischen Gleichung mit nichtkonformen finiten Elementen*, Diplomarbeit, Universität Heidelberg, 1992

[6] Schwarz, H.R.: *Methode der finiten Elemente*, Teubner, Stuttgart 1984

[7] Turek, S.: *Ein robustes und effizientes Mehrgitterverfahren zur Lösung der instationären, inkompressiblen 2-D Navier–Stokes–Gleichungen mit diskret divergenzfreien finiten Elementen*, Dissertation, Heidelberg 1991

[8] Turek, S.: *Mehrgitterverfahren für eine Klasse von nichtkonformen Finite Elemente Diskretisierungen*, Diplomarbeit, Universität Saarbrücken, 1988

[9] Turek, S.: *A multigrid Stokes solver using divergence free finite elements*, Technischer Report Nr. 527, SFB 123, Universität Heidelberg, Juli 1989

The effect of incomplete decomposition preconditioning on the convergence of Conjugate Gradients

Henk A. Van der Vorst *   Gerard G. L. Sleijpen*

1. Introduction

The ICCG method, for the iterative solution of the large sparse linear system $Ax = b$ (A is symmetric positive definite) consists of the preconditioning of the system, using an incomplete Choleski decomposition K of the matrix A, and the subsequent iterative solution of this system by the conjugate gradients method. For the construction of an incomplete Choleski decomposition it is, in addition, desirable that A be an M-matrix (see, e.g., [12]).

If A is not an M-matrix then it may be necessary to take measures in order to prevent the incomplete decomposition algorithm from failing or from leading to unstable decompositions.

The improvement of the convergence behaviour of the preconditioned conjugate gradient method, as observed in practice, can only be partly understood from a reduction of the condition number. A favourable distribution of the eigenvalues appears to be an important factor. For a simple model problem we shall derive approximations for relevant eigenvalues of the preconditioned matrix $K^{-1}A$ that will help to explain the convergence behaviour.

It should be mentioned that the model problem has two disadvantages: such problems can be solved much more efficiently by certain other methods (e.g., Fast Poisson Solvers or Multigrid), and in more realistic cases (e.g., with strongly varying coefficients) often a much more impressive improvement of the convergence behaviour is observed [6].

In this paper we will first give the algorithm for conjugate gradients, together with some of the main properties of the method. Then we will for a model problem analyse the effects of preconditioning. Throughout this paper we will assume exact arithmetic.

2. The conjugate gradient method

For the solution of the linear system $Ax = b$, with A a symmetric and positive definite n by n matrix, the conjugate gradient (CG) method can be used. This method may be

---

*Mathematical Institute, University of Utrecht, P.O. Box 80.010, NL-3508 TA Utrecht, the Netherlands,

represented in various alternative forms, of which we have selected the following one:

$x_0=$ initial guess; $r_0 = b - Ax_0$;
$p_{-1} = 0; \beta_{-1} = 0$;
$\rho_0 = (r_0, r_0)$
for $i = 0, 1, 2, ...$
$\quad p_i = r_i + \beta_{i-1} p_{i-1}$;
$\quad q_i = Ap_i$;
$\quad \alpha_i = \frac{\rho_i}{(p_i, q_i)}$
$\quad x_{i+1} = x_i + \alpha_i p_i$;
$\quad r_{i+1} = r_i - \alpha_i q_i$;
$\quad$ if x_{i+1} accurate enough then quit;
$\quad \rho_{i+1} = (r_{i+1}, r_{i+1})$;
$\quad \beta_i = \frac{\rho_{i+1}}{\rho_i}$;
end;

For a discussion on the CG method and its main properties we refer to [2], [3], [4] and [9]. We recall the following properties:

1. The CG method minimizes $\|x_i - x\|_A \equiv (x_i - x, A(x_i - x))^{\frac{1}{2}}$ for all i, among all algorithms of the form

$$x_i = x_0 + Q_{i-1}(A)A(x - x_0), \tag{1}$$

where $Q_{i-1}(A)$ is an $(i - 1)$-st degree polynomial in A.
Note: many well-known iteration methods can be written in this form, e.g., Richardson iteration, Chebychev iteration.

2. Iteration (1) can be rewritten as

$$x_i - x = (I - AQ_{i-1}(A))(x_0 - x) = P_i(A)(x_0 - x), \tag{2}$$

with $P_i(0) = 1$. If we exploit the optimality property, then by selecting for P_i a shifted and normalized Chebyshev polynomial, we obtain the well-known upperbound for the error in A-norm ([1],[3]):

$$\|x_i - x\|_A \leq 2 \left(\frac{\sqrt{\kappa_2} - 1}{\sqrt{\kappa_2} + 1} \right)^i \|x_0 - x\|_A, \tag{3}$$

where κ_2 is the condition number of A with respect to the 2-norm. This upperbound is often quite realistic in the early phase of the iteration process, but it turns out to be very pessimistic on the long run since it does not account for the so-called superlinear convergence behaviour [2], [9].

3. The vectors r_0, ..., r_i are mutually orthogonal and they satisfy the recurrence relation

$$Ar_j = -\frac{\beta_{j-1}}{\alpha_{j-1}} r_{j-1} + \left(\frac{1}{\alpha_j} + \frac{\beta_{j-1}}{\alpha_{j-1}} \right) r_j - \frac{1}{\alpha_j} r_{j+1}, \tag{4}$$

which describes essentially the Lanczos algorithm [8].
Let R_i denote the matrix with columns r_0, ..., r_i, then the recurrence relation can

be rewritten as

$$AR_i = R_i T_i - \frac{1}{\alpha_i} r_{i+1} e_{i+1}^T, \tag{5}$$

where $e_{i+1}^T = (0, .., 0, 1)$, the $(i+1)$-st unit vector of dimension $i+1$, and T_i is a tridiagonal matrix of order $i+1$.

4. The eigenvalues of T_i are called the Ritz values of A with respect to the subspace spanned by r_0, ..., r_i. These Ritz values approximate some eigenvalues of A increasingly well for increasing i. It has been observed that when an extremal eigenvalue has been sufficiently well approximated by a Ritz value, then the CG process converges faster from then on (superlinear convergence behaviour) [3]. This was further analyzed and quantified in [9].

5. More refined errorbounds for CG are given in [1] and [5]. These bounds take advantage from the distribution of the (extremal) eigenvalues.

CG is most often used in combination with a suitable splitting $A = K - R$, and then K^{-1} is called the preconditioner. We will assume that K is also positive definite.
A popular scheme that includes preconditioning can be obtained from the previous scheme as follows. First we remark that the CG method can be derived for any choice of the innerproduct. If we make the choice:

$$[x, y] \equiv (x, Ky),$$

where $(\ ,\ )$ denotes the innerproduct: $(x, y) = \sum x_i y_i$, then it is easy to verify that $K^{-1}A$ is symmetric positive definite with respect to $[\ ,\ ]$:

$$[K^{-1}Ax, y] = (K^{-1}Ax, Ky) = (Ax, y) = (x, Ay) = [x, K^{-1}Ay].$$

Hence, we can formulate the CG procedure for solving the preconditioned system $K^{-1}Ax = K^{-1}b$, using the new $[\ ,\ ]$-innerproduct.
Apparently, we are then minimizing

$$[x_i - x, K^{-1}A(x_i - x)] = (x_i - x, A(x_i - x)),$$

which leads to the remarkable result that for the preconditioned system we still minimize the error in A-norm, but now with A replaced by $K^{-1}A$ in (1).

In the following computational scheme for preconditioned CG, for the solution of $Ax = b$ with preconditioner K^{-1}, we have replaced the $[\ ,\ ]$-innerproduct again by the familiar standard innerproduct. E.g., note that

$$\rho_{i+1} = [K^{-1}r_{i+1}, K^{-1}r_{i+1}] = (r_{i+1}, K^{-1}r_{i+1}),$$

and $K^{-1}r_{i+1}$ is the residual corresponding to the preconditioned system

$$K^{-1}Ax = K^{-1}b.$$

$x_0 =$ initial guess; $r_0 = b - Ax_0$;
$p_{-1} = 0; \beta_{-1} = 0;$

Solve w_0 from $Kw_0 = r_0$;
$\rho_0 = (r_0, w_0)$
for $i = 0, 1, 2, \ldots.$
$\quad p_i = w_i + \beta_{i-1}p_{i-1}$;
$\quad q_i = Ap_i$;
$\quad \alpha_i = \frac{\rho_i}{(p_i, q_i)}$
$\quad x_{i+1} = x_i + \alpha_i p_i$;
$\quad r_{i+1} = r_i - \alpha_i q_i$;
$\quad$ if x_{i+1} accurate enough then quit;
$\quad$ Solve w_{i+1} from $Kw_{i+1} = r_{i+1}$;
$\quad \rho_{i+1} = (r_{i+1}, w_{i+1})$;
$\quad \beta_i = \frac{\rho_{i+1}}{\rho_i}$;
end;

Note that this formulation, which is quite popular, has the advantage that the preconditioner needs not to be split into two factors, and it is also avoided to backtransform solutions and residuals, as is necessary when one applies CG to $L^{-1}AL^{-1^T}y = L^{-1}b$ (for $K = LL^T$).

3. The model problem

We consider the Poisson equation
$$-\Delta u = f, \tag{6}$$
over the unit square in $I\!R^2$, with Dirichlet boundary conditions. Standard 5-point finite difference discretization with stepsize $\frac{1}{m+1}$ in both directions will be used, which leads to a linear system with m^2 unknowns and with a matrix A, as given in (7). An equivalent analysis can be done for unequal stepsizes and this leads to essentially the same results, but it makes most of the formulas less transparant.

$$A = \begin{pmatrix} A_1 & C_1 & & & & & \\ C_1 & A_2 & C_2 & & & & \\ & C_2 & A_3 & \ddots & & & \\ & & \ddots & \ddots & \ddots & & \\ & & & \ddots & \ddots & \ddots & \\ & & & & \ddots & \ddots & C_{m-1} \\ & & & & & C_{m-1} & A_m \end{pmatrix} \tag{7}$$

with

$$
A_j = \begin{pmatrix}
a_{1,j} & b_{1,j} \\
b_{1,j} & a_{2,j} & b_{2,j} \\
& b_{2,j} & a_{3,j} & \ddots \\
& & \ddots & \ddots & \ddots \\
& & & \ddots & \ddots & \ddots \\
& & & & \ddots & \ddots & b_{m-1,j} \\
& & & & & b_{m-1,j} & a_{m,j}
\end{pmatrix},
\tag{8}
$$

$$
C_j = \begin{pmatrix}
c_{1,j} \\
& c_{2,j} \\
& & \ddots \\
& & & \ddots \\
& & & & c_{m,j}
\end{pmatrix}.
\tag{9}
$$

The values for $a_{i,j}$, $b_{i,j}$ and $c_{i,j}$ are: $a_{i,j} = 4$, $b_{i,j} = -1$, $c_{i,j} = -1$.
The eigenvectors of A are given by

$$
v_{i,j}^{k,\ell} = \sin \frac{ik\pi}{m+1} \sin \frac{j\ell\pi}{m+1}, \quad 1 \leq i,j,k,\ell \leq m,
\tag{10}
$$

where $v_{i,j}^{k,\ell}$ is the (i,j)-th component of $v^{k,\ell}$, corresponding to the i-th gridpoint on the j-th gridline (there are m gridlines with m gridpoints each).
The eigenvalue corresponding to $v^{k,\ell}$ is

$$
\lambda^{k,\ell} = 4 - 2\cos \frac{\pi k}{m+1} - 2\cos \frac{\pi \ell}{m+1}.
\tag{11}
$$

Throughout this paper the incomplete decomposition of A will be the following splitting:

$$
A = K - R \quad \text{with} \quad K = (L + \tilde{D})\tilde{D}^{-1}(\tilde{D} + L^T),
\tag{12}
$$

where K is defined by

1. $\mathrm{diag}\left((L + \tilde{D})\tilde{D}^{-1}(\tilde{D} + L^T)\right) = \mathrm{diag}(A) \equiv D$

2. L is a lower triangular matrix, whose subdiagonal elements equal those of A

3. $\tilde{D}$ is a diagonal matrix whose diagonal elements equal those of L.

If, according to the notation in (8), the diagonal elements of $\tilde{D}$ are denoted by $d_{i,j}$, then they can be computed recursively from

$$
d_{i,j} = a_{i,j} - \frac{b_{i-1,j}^2}{d_{i-1,j}} - \frac{c_{i,j-1}^2}{d_{i,j-1}}
\tag{13}
$$

(<u>Convention</u>: Here, and elsewhere in this paper, terms involving factors with an index 0 or $m+1$ should be omitted).

For any vector u we have

$$(Ru)_{i,j} = \frac{b_{i,j-1}c_{i,j-1}}{d_{i,j-1}}u_{i+1,j-1} + \frac{b_{i-1,j}c_{i-1,j}}{d_{i-1,j}}u_{i-1,j+1}. \tag{14}$$

This incomplete decomposition belongs to the class of incomplete Choleski decompositions [7]; it is the standard decomposition with no fill-in. The existence of these and other incomplete decompositions was proven for M-matrices in [7].

For the model problem it is straightforward to prove from the recurrence relations (13) that

$$0 < \frac{1}{d_{i,j}} < \frac{1}{d} \quad \text{with} \quad d = 2 + \sqrt{2}. \tag{15}$$

4. Eigenvalues of the preconditioned matrix

For the model problem (the discretized Poisson problem) we can derive quite accurate bounds on the eigenvalues of the preconditioned matrix. In this section we will discuss the bound for the largest eigenvalue only, the derivation for the bounds on the smallest eigenvalues follows similar ideas and will be omitted here.

The preconditioned matrix $K^{-1}A$ is a product of two symmetric positive definite matrices and hence its eigenvalues are positive real. The splitting $A = K - R$, as defined above, defines a regular splitting [7], and therefore the spectrum of $K^{-1}A$ is contained in the interval $(0,2)$. This result guarantees the convergence of the basic iteration $x_i = x_{i-1} + K^{-1}(b - Ax_{i-1})$, but it does not explain the success of preconditioning in the conjugate gradient process. To that end we need more refined bounds.

Note that if λ is an eigenvalue of $K^{-1}A$, then 0 must be in the spectrum of $\lambda K - A$ (denoted by $\sigma(\lambda K - A)$). The main idea is to split from $\lambda K - A$ a positive definite part as large as possible. This will help us to bound $\sigma(\lambda K - A)$ from below. It is easy to verify that

$$\lambda K - A = \frac{1}{\lambda}\left[((1-\lambda)\tilde{D} - \lambda L)\tilde{D}^{-1}((1-\lambda)\tilde{D} - \lambda L^T) - (1-2\lambda)\tilde{D} - \lambda D\right]$$

$$= \frac{1}{\lambda}\left[W(\lambda) + V(\lambda)\right] \tag{16}$$

with $V(\lambda) = -(1 - 2\lambda)\tilde{D} - \lambda D$. Since $W(\lambda)$ is symmetric non-negative definite, the diagonal matrix $V(\lambda)$ cannot be positive definite if we want to have $\lambda K - A$ singular. For our model problem this leads immediately to the condition that

$$\lambda \le \frac{d}{2d-4} = (1 + \sqrt{2})/2 \approx 1.20711. \tag{17}$$

For the 30 by 30 model problem the true largest eigenvalue of the preconditioned matrix has been computed to be ≈ 1.20455. This is remarkably close to the derived upperbound.

By slightly more complicated techniques also bounds for some of the smallest eigenvalues of $K^{-1}A$ can be derived (this will be published elsewhere). These eigenvalue approximations can be used to bound the number of preconditioned CG steps necessary to reduce the residual by at least 10^{-12}. To that goal we used Chebychev-type bounds, as proposed in [5]. This approach led to a prediction of 51 iterations at most, if we used only the information for the five smallest eigenvalues and the upperbound for the largest one [11]. The position of all the other eigenvalues is irrelevant for obtaining that bound on the number of iterationsteps. Would we have used the exact locations of the required eigenvalues, then that would have led to a predicted 50 iterationsteps. In actual computation we have seen as many as 44 iterationsteps for some random generated right-hand sides.

We see that only the largest eigenvalue as well as the position of a few of the smallest eigenvalues is really necessary for getting a realistic upperbound on the number of iterationsteps (at least for the required reduction factor for the residuals). Here we have taken the 30 by 30 model problem as an example. In the next section we will discuss the convergence behaviour of CG for more general sizes.

5. A discussion on the convergence of ICCG

In the previous section we have argued that at least for the 30 by 30 model problem the number of iterationsteps for preconditioned CG could be predicted quite accurately. A natural question is, of course, whether this situation is typical for what happens when the size (m) is much larger. In this section we will shed light on this matter.
For m large, the bounds on the smallest eigenvalues, as sketched in the previous section, turn out to be rather close. Hence, denoting the eigenvalues of $K^{-1}A$ as $\lambda_1 \leq \lambda_2 \leq \lambda_3 \leq \dots$, we have for an increasing number of successive smallest eigenvalues (as m increases):

$$\lambda_1 \approx \frac{d}{2}\lambda^{1,1}, \lambda_2 \approx \lambda_3 \approx \frac{d}{2}\lambda^{1,2} = \frac{d}{2}\lambda^{2,1}, \lambda_4 \approx \frac{d}{2}\lambda^{2,2}, \dots \tag{18}$$

Since $\lambda_{max}(A) \approx 8$ and $\lambda_{max}(K^{-1}A) \approx 1.207$, we have for the condition number $\kappa_2(A)$ of A

$$\kappa_2(A) \approx \frac{4(m+1)^2}{\pi^2}, \tag{19}$$

and for the condition number of $K^{-1}A$ we find

$$\kappa_2(K^{-1}A) \approx \frac{1.207}{4(2+\sqrt{2})}. \tag{20}$$

The rates of convergence of the CG algorithm for the unpreconditioned and the preconditioned case are then, according to (3)

$$r_A = \frac{\sqrt{\kappa_2(A)} - 1}{\sqrt{\kappa_2(A)} + 1} \approx 1 - \frac{2}{\sqrt{\kappa_2(A)}}, \tag{21}$$

and

$$r_{K^{-1}A} = \frac{\sqrt{\kappa_2(A)/\alpha} - 1}{\sqrt{\kappa_2(A)/\alpha} + 1} \approx 1 - \frac{2}{\sqrt{\kappa_2(A)}}\sqrt{\alpha}, \tag{22}$$

with $\sqrt{\alpha} \approx 3.3$.

We will further refer to the unpreconditioned process as CG and to the preconditioned process as ICCG.

We conclude that in the first phase of the iteration process, i.e., the phase in which the convergence behaviour can be described reasonably well by the condition number, a comparative accuracy will be achieved by ICCG within $\approx \sqrt{\alpha}$ times fewer iterations, and we note that this factor is independent of m. This corresponds quite nicely to what we see for the actual numbers of iterations when only moderate accuracies are required (say 10^{-6} or less).

Once the smallest Ritz value, associated with CG, has come close to the smallest eigenvalue of A (and even a modest degree of approximation suffices, see [9]), the convergence behaviour of CG will be more appropriately described by the "effective" condition number $8/\lambda^{1,2}$. Likewise, as soon as this happens for ICCG, the effective condition number for ICCG will be at most $1/\alpha$ times that value. Now we note that the length of the spectrum of $K^{-1}A$ is smaller than the length of the spectrum of A, and on account of the estimates for the smallest eigenvalues we have

$$|\lambda_1 - \lambda_2| \approx \frac{d}{2}|\lambda^{1,1} - \lambda^{1,2}| \tag{23}$$

with $\frac{d}{2} > 1$).

Hence, in the terminology of Parlett [8], the smallest eigenvalue of $K^{-1}A$ has a larger gap ratio than the corresponding one of A. In addition the smallest eigenvalues of A and $K^{-1}A$ share almost the same eigenvectors (for m not too small), see [11]. Theorem 12-4-1 in [8] then indicates that we may expect the smallest eigenvalue of $K^{-1}A$ to be approximated to the same accuracy within fewer iterations than is the case for A.

Therefore we may expect ICCG to take sooner advantage of a reduced effective condition number (and, indeed: for $m = 30$ we find in practice that the first Ritz value for ICCG converges to 5 decimal places, within the first 15 iterations, whereas for CG this takes about 50 iterations).

A similar discussion about the subsequent phases might seem more complicated, because of the occurrence of almost double eigenvalues in ICCG, which leads to complications in the convergence of the Ritz values towards these close eigenvalues [10]. However, in [9] it has been shown that an almost double eigenvalue leads to only a modest delay in convergence of the (preconditioned) CG process, compared to the situation where the almost double eigenvalues have been replaced by one single eigenvalue.

Here we have focussed on the separation of the smallest eigenvalues. Note that we need much larger gaps at the upper end of the spectrum in order to obtain similar reductions in the effective condition number.

In conclusion we see that the successive phases of the iteration process for ICCG are entered within fewer iterations than the corresponding ones in CG. Since, in addition, in each phase of ICCG the convergence is $\approx \sqrt{\alpha}$ times faster than the corresponding phase of CG, it is clear that we should expect ICCG to be very much faster than CG for large values of m. Note that it is, contrary to popular belief, not so much the clustering of the eigenvalues near 1 which helps to explain the effect of preconditioning, but that this effect, at least for practical reductions of the error, can be largely explained by the reduction of the condition number and the improved relative separation of only a few of the smallest eigenvalues.

References

[1] O. Axelsson. Solution of linear systems of equations: iterative methods. In V. A. Barker, editor, *Sparse Matrix Techniques*, Copenhagen 1976, Springer Verlag, Berlin, 1977.

[2] P. Concus, G. H. Golub, and D. P. O'Leary. A generalized conjugate gradient method for the numerical solution of elliptic partial differential equations. In J. R. Bunch and D. J. Rose, editors, *Sparse Matrix Computations*, Academic Press, New York, 1976.

[3] G. H. Golub and C. F. van Loan. *Matrix Computations*. The Johns Hopkins University Press, Baltimore, 1989.

[4] M. R. Hestenes and E. Stiefel. Methods of conjugate gradients for solving linear systems. *J. Res. Natl. Bur. Stand.*, 49:409–436, 1954.

[5] A. Jennings. Influence of the Eigenvalue Spectrum on the Convergence Rate of the Conjugate Gradient Method. *J. Inst. Math. Applic.*, 20:61-72, 1977.

[6] D. S. Kershaw. The Incomplete Cholesky - Conjugate Gradient method for the iterative solution of systems of linear equations. *J. of Comp. Phys.*, 26:43–65, 1978.

[7] J. A. Meijerink and H. A. van der Vorst. An iterative solution method for linear systems of which the coefficient matrix is a symmetric M-matrix. *Math.Comp.*, 31:148–162, 1977.

[8] B. N. Parlett. *The Symmetric Eigenvalue Problem*. Prentice-Hall, Englewood Cliffs, N.J., 1980.

[9] A. van der Sluis and H. A. van der Vorst. The rate of convergence of conjugate gradients. *Numer. Math.*, 48:543–560, 1986.

[10] A. van der Sluis and H. A. van der Vorst. The convergence behavior of Ritz values in the presence of close eigenvalues. *Lin. Alg. and its Appl.*, 88/89:651–694, 1987.

[11] H. A. van der Vorst. Preconditioning by Incomplete Decompositions. PhD Thesis, Utrecht University, 1982.

[12] R. S. Varga. *Matrix Iterative Analysis*. Prentice-Hall, Englewood Cliffs N.J., 1962.

On Parallel Frequency Filtering

W. Weiler †, G. Wittum
SFB 123, Universität Heidelberg
Im Neuenheimer Feld 294, D-6900 Heidelberg

Summary

The frequency filtering method is a robust and efficient ILU-like solver for large sparse systems (cf. [Wi1,2]). Combining this method with the so-called Schur-complement DD method we obtain a fast parallel solver. In this context frequency filtering can be applied as solver inside the subdomains as well as for the treatment of the arising Schur-complements. Especially for those the method is well suited since it is highly parallelizable by recursively applying the same decomposition as to the original system. In this paper an implementation of the frequency filtering domain decomposition (FFDD) method on a multiprocessor system will be presented and the numerical results of some variants thereof will be discussed. The scaling behaviour of the algorithm for an increasing number of processors is almost optimal.

1. Introduction

In the present paper we consider the elliptic diffusion-type equation

$$-\text{div}(\phi(x,y)\,\nabla u) = f \qquad \text{in } \Omega = (0,1)\times(0,1) \tag{1.1}$$
$$u = u_{Rand} \quad \text{on } \partial\Omega$$

with $C_1 \geq \phi(x,y) \geq C_2 > 0$ as model problem. Discretizing (1.1) on an equidistant cartesian grid Ω_h by a symmetric box method and numbering the grid points lexicographically, we obtain a block-tridiagonal stiffness matrix

$$K = \begin{pmatrix} D_1 & L_1^T & & \\ L_1 & D_2 & \ddots & \\ & \ddots & \ddots & L_{m-1}^T \\ & & L_{m-1} & D_m \end{pmatrix} \tag{1.2}$$

with $n \times n$ blocks L_i, D_i, where the D_i are symmetric. K is sparse, symmetric and positive definite. We then have to solve the linear system

This work has been supported by Deutsche Forschungsgemeinschaft

$$K\,x\ =\ b\,. \tag{1.3}$$

We consider a linear iterative method for solving (1.3)

$$x^{new}\ =\ x^{old} + M^{-1}\,(b - K\,x^{old}) \tag{1.4}$$

constructed by splitting

$$K\ =\ M - N \tag{1.5}$$

where M is regular and has the form

$$M\ =\ (L+T)\,T^{-1}(L^{T}+T) \tag{1.6}$$

with a lower triangular matrix L and

$$T\ =\ \text{blockdiag}\,\{T_{i},\, i\ =\ 1,\dots,n\,\}. \tag{1.7}$$

2. Frequency Filtering

2.1 Frequency Filtering Decompositions

A frequency filtering decomposition of order p for K from (1.2) is constructed such that the resulting approximate inverse M from (1.6) has the same effect as K on the probe vectors

$$\mathbf{e}_{j}\ =\ (\sin(v_{j}\pi i h))_{i=1,\dots,n}. \tag{2.1}$$

The entries of the block-diagonal matrix T from (1.7) are computed by the recursion formula

$$T_{1}\ =\ D_{1} \tag{2.2}$$

$$T_{i}\ =\ D_{i} - \Theta_{i}\,,\, 1 \le i \le n \tag{2.3}$$

with a symmetric $(2p+1)$-diagonal matrix Θ_{i} determined by the filtering property

$$\Theta_{i}\,\mathbf{e}_{j}\ =\ (L_{i-1}T_{i-1}^{-1} L_{i-1}^{T})\,\mathbf{e}_{j}\,,\, j\ =\ 0,\dots,p\,, \tag{2.4}$$

for a certain set of test frequencies $v_{0},\dots,v_{p}$. For the above choice of testing vectors $\mathbf{e}_{j}$ the Θ_{i} can uniquely be determined from (2.3), see [Wi1,2]. In the present paper we restrict ourselves to $p = 0$ (Θ_{i} diagonal) and $p = 1$ (Θ_{i} tridiagonal). Since the main diagonal blocks of K are tridiagonal, the sparsity pattern is preseverd by this choice of p and the resulting T_{i} will also be tridiagonal.

Remark 2.1: *The filtering property (2.3) implies that*

$$M\mathbf{z}\ =\ K\mathbf{z},\ \textit{for all } \mathbf{z} \in \{\mathbf{e}_{j} \otimes \mathbf{y}\} \tag{2.5}$$

$\otimes$ denoting the tensor product defined by

$$(\mathfrak{a} \otimes \mathfrak{b})_{i \cdot n + j} := \mathfrak{a}_i \mathfrak{b}_j, \quad \mathfrak{a}, \mathfrak{b} \in \mathbb{R}^n, \mathfrak{a} \otimes \mathfrak{b} \in \mathbb{R}^{n \cdot n}. \tag{2.6}$$

For small frequencies v the probe vectors are smooth and the iteration matrix

$$S_v = I - M_v^{-1} K \tag{2.7}$$

for the linear iteration (1.4) is exact on a smooth subspace and thus S_v can be expected to be a good corrector. Choosing high testing frequencies, the probe vectors are rough and S_v will be a good smoother.

2.2. The FF-Solver

The idea of the frequency filtering (FF-) solver is to combine frequency filtering decompositions acting as smoother and corrector. To that purpose we choose a logarithmic sequence of testing frequencies which for $p = 1$ is given by

$$v_0^{(1)} = 1,$$

$$v_0^{(i+1)} = \begin{cases} \left[\alpha v_0^{(i)} \right] & \text{if } \left[\alpha v_0^{(i)} \right] > v_1^{(i)} \\ v_1^{(i)} + 1 & \text{otherwise} \end{cases} \tag{2.8}$$

$$v_1^{(i)} = v_0^{(i)} + 1.$$

with $\alpha > 1$. In the present paper we choose $\alpha = 2$. For each pair of testing frequencies $v^{(i)} = (v_0^{(i)}, v_1^{(i)})$ we obtain a frequency filtering decomposition $M_{v^{(i)}}$ and a corresponding iteration operator $S_{v^{(i)}}$. Applying these operators consecutively we get a product iteration with the operator

$$S_\Pi = \prod_{i=1}^{\log_\alpha n} S_{v^{(i)}} . \tag{2.9}$$

In [Wi1,2] it has been proved that if $\phi(x,y) = \text{const}$ in (1.1) this product iteration has a convergence rate which is independent of the size of the system, i.e.

$$\| S_\Pi \| \leq \zeta < 1 \tag{2.10}$$

holds in the Euclidean norm independently on h. Various numerical tests in [Wi1,2] show that the FF-solver also works well for problems with various coefficients, nonsymmetric and nonlinear problems.

3. Parallel Frequency Filtering

3.1 The capacitance system

The method descibed above is highly recursive. To parallelize it we combine it with the so-called Schur-complement DD method. To that purpose we subdivide the grid Ω_h into p stripes $\Omega_{h,i}$, $i = 1, \dots, p$, with $\Omega_h = \cup_{i=1}^{p}\Omega_{h,i}$, $\Omega_{h,i}\cap\Omega_{h,j} = \varnothing$, $i \neq j$ with interfaces $\Gamma_{h,i}$, $i = 1,\dots,p-1$, as shown in Figure 1. Now we partition K as usual, taking first all unknowns inside the stripes and thereafter the unknowns on the interfaces $\Gamma_{h,i}$. Then system (1.3) has the form

$$\begin{pmatrix} A & B \\ B^T & C \end{pmatrix}\begin{pmatrix} x_\Omega \\ x_\Gamma \end{pmatrix} = \begin{pmatrix} f_\Omega \\ f_\Gamma \end{pmatrix} \tag{3.1}$$

where the subscripts Ω and Γ denote the corresponding variables on the grids $\Omega_{h,i}$ and $\Gamma_{h,i}$. The matrix A has block-diagonal form

$$A = \text{blockdiag}\,\{A_i, 1 \leq i \leq p\} \tag{3.2}$$

with $n \cdot m \times n \cdot m$ blocks A_i of the same form as K in (1.2), representing the discretization of (1.1) on the strip Ω_i with zero boundary condition on Γ_i.

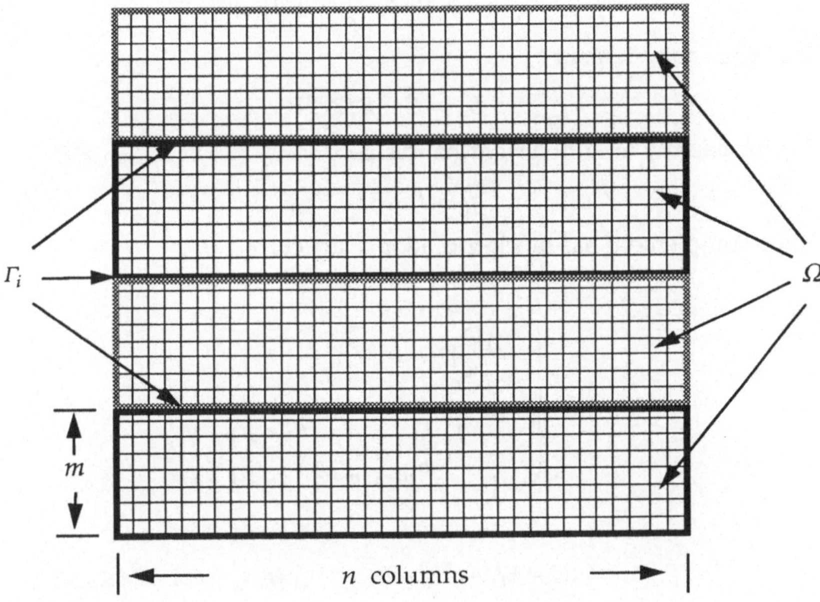

Figure 1: *Partitioning of the grid Ω_h into stripes. The stripes $\Omega_{h,i}$ and the discrete interfaces $\Gamma_{h,i}$ are indicated by arrows.*

The coupling block B has the form

$$B = \begin{pmatrix} B_{11} & & & & \\ B_{21} & B_{22} & & & \\ & \ddots & & \ddots & \\ & & B_{p-1,p-2} & B_{p-1,p-1} & \\ & & & B_{p,p-1} & \end{pmatrix}. \tag{3.3}$$

The block B_{ij} of size $n \cdot m \times n$ represents the coupling of the unknowns in $\Omega_{h,i}$ with the nodes on the interface $\Gamma_{h,i}$. Since the 5-point stencil couples only the first and last row of the subdomain $\Omega_{h,i}$ with the corresponding interfaces $\Gamma_{h,i-1}$ and $\Gamma_{h,i}$, B_{ij} has non-zero entries only in the first and last $n \times n$ block.

The matrix C is block-diagonal, $C = \text{blockdiag}\{C_i, 1 \le i \le p-1\}$, with $n \times n$ blocks C_i representing the discretized operator on the internal boundary Γ_i.

Block elimination of system (3.1) leads to

$$\begin{pmatrix} A & B \\ 0 & S \end{pmatrix} \begin{pmatrix} x_\Omega \\ x_\Gamma \end{pmatrix} = \begin{pmatrix} f_\Omega \\ \tilde{f}_\Gamma \end{pmatrix} \tag{3.4}$$

with the Schur complement

$$S = C - B^T A^{-1} B \tag{3.5}$$

and the consistently modified right-hand side

$$\tilde{f}_\Gamma = f_\Gamma - B^T A^{-1} f_\Omega . \tag{3.6}$$

The Schur-complement has block-tridiagonal structure

$$S = \begin{pmatrix} S_{11} & S_{12} & & & \\ S_{21} & S_{22} & & \ddots & \\ & \ddots & \ddots & & S_{p-2\ p-1} \\ & & S_{p-1\ p-2} & S_{p-1\ p-1} & \end{pmatrix} \tag{3.7}$$

with

$$\begin{aligned}
S_{i,i} &= C_i - B_{ii}^T A_i^{-1} B_{ii} - B_{i+1,i}^T A_{i+1}^{-1} B_{i+1,i} && \text{for } 1 \le i \le p-1, && (3.8) \\
S_{i,i+1} &= -B_{i+1,i}^T A_{i+1}^{-1} B_{i+1,i+1} && \text{for } 1 \le i \le p-2, && (3.9) \\
S_{i,i-1} &= -B_{ii}^T A_i^{-1} B_{ii-1} && \text{for } 2 \le i \le p-1. && (3.10)
\end{aligned}$$

In the original substructuring approach (cf. [BW]) the reduced system (3.4) was solved directly. This is not realistic if we think of large systems and using iterative solvers for the inner solves. Instead we construct an approximate inverse for system (3.1).

3.2 The FFDD Algorithm

We want to solve system (3.1) with a globally applied FF-solver. Our approximate inverse M has the form

$$M_\nu = \begin{pmatrix} I & \\ B^T A_{FF_\nu}^{-1} & I \end{pmatrix} \begin{pmatrix} A_{FF_\nu} & \\ & S_{FF_\nu} \end{pmatrix} \begin{pmatrix} I & A_{FF_\nu}^{-1} B \\ & I \end{pmatrix} \tag{3.11}$$

where $A_{FF_\nu} = \text{blockdiag}\{A_{1,FF_\nu} \cdots, A_{p,FF_\nu}\}$ denotes the frequency filtering decomposition on the internal subdomains Ω_i. S_{FF_ν} is a filtering decompostion approximating the Schur-complement S the construction of which is described in the following.

The problem in finding an approximate inverse for S is mainly caused by the fact that the blocks are full. With the filtering technique descibed above, however, we have a powerful tool to construct sparse approximations to full matrices such as Schur-complements. This means that we first construct a sparse approximaton $\tilde{S}_{FF_\nu}$ to S using frequency filtering techniques which is decomposed thereafter into a FF-decomposition S_{FF_ν}.

We tried two different variants to construct $\tilde{S}_{FF_\nu}$. In the first case $\tilde{S}_{FF_\nu}$ was chosen to be blockdiagonal:

$$\tilde{S}_{FF_\nu} = \text{blockdiag}\{\tilde{S}_{1,FF_\nu} \cdots, \tilde{S}_{p-1,FF_\nu}\} \tag{3.12}$$

with tridiagonal blocks $\tilde{S}_{i,FF_\nu}$ having the same effect on the testing vectors $\mathbf{e}_j$ from (2.1)

$$\tilde{S}_{i,FF_\nu} \mathbf{e}_j = \tilde{S}_{ii} \mathbf{e}_j, \text{ for } j = 0,1. \tag{3.13}$$

This corresponds to the Dryija type preconditioner, [Dr], choosing a one-dimensional operator which acts only on the interface nodes and neglects the coupling between neighboring interfaces. The condition number of Dryija's preconditioner is known to deteriorate like $1/p$ with an increasing number of stripes (see [Dr, Ha1-3]). This is due to neglecting the couplings.

As a second choice we used a block-tridiagonal approximation

$$\tilde{S}_{FF_\nu} = \begin{pmatrix} \tilde{S}_{11,FF_\nu} & \tilde{S}_{12,FF_\nu} & & & \\ \tilde{S}_{21,FF_\nu} & \tilde{S}_{22,FF_\nu} & \ddots & & \\ & \ddots & \ddots & \tilde{S}_{p-2\,p-1,FF_\nu} \\ & & \tilde{S}_{p-1\,p-2,FF_\nu} & \tilde{S}_{p-1\,p-1,FF_\nu} \end{pmatrix} \tag{3.14}$$

with tridiagonal blocks $\tilde{S}_{ii,FF_v}$ satisfying

$$\tilde{S}_{ii,FF_v}\mathbf{e}_j = S_{ii}\,\mathbf{e}_j\,,\ \text{for}\ j = 0,1. \tag{3.15}$$

For the off-diagonal blocks we use either diagonal ($p=0$) or tridiagonal ($p=1$) approximations :

$$\tilde{S}_{il,FF_v}\mathbf{e}_j = S_{il}\,\mathbf{e}_j\,,\ \text{for}\ j = 0,p\,,\ i \neq l. \tag{3.16}$$

The approximations to the off-diagonal blocks of the Schur-complement represent the coupling of neighboring internal boundaries with each other. Thus we obtain a sparse approximation of the Schur-complement which is exact on the subspace treated by the corresponding iteration operator.

The decomposition is computed performing the following three steps

1. Compute the FF-decomposition A_{i,FF_v} for the inner matrices A_i.
2. Replace the blocks S_{il} by approximations $\tilde{S}_{il,FF_v}$ satisfying (3.14).
   On account of (2.3) the action of the Schur-complement on the testing vectors $\mathbf{e}_j$ can easily be calculated exactly using the FF-decompositions from step 1.
3. Compute the FF-decomposition of the approximation S_{FF_v}.

3.3 Recursive Partitioning

Since S has the same block structure as our original matrix K, it can be split up in the same way as K. This means we consider S as a system on a domain made up from $p-1$ gridlines $\Gamma_{i,h}$. Now we split up this system again into q clusters $G_l = \{\Gamma_{i_{k-1}+1},\ldots,\Gamma_{i_k-1}\}$ and $q-1$ interfaces Γ_{i_k} for which we can construct a new Schur-complement. Thus we get a recursive procedure $\tilde{S}_{FF_v}(K^{(i)})$ constructing approximations of the Schur-complement of a given system $K^{(i)}$.

$$K^{(maxlevel)} := K \tag{3.17}$$
$$K^{(i-1)} := \tilde{S}_{FF_v}(K^{(i)})\ \text{for}\ maxlevel > i > 0.$$

As for the FF-solver we choose for the frequency filtering domain decomposition (FFDD) algorithm a logarithmic sequence of test frequencies $v_0^{(i)}$ and $v_1^{(i)}$. The FFDD-algorithm then looks like a multigrid algorithm, reducing the problem of inverting the Schur-complement in a tree-like manner from finer downto coarser levels. For a given pair of test frequencies (v_0, v_1) it reads

$$FFDD_v\_setup(K,l) \tag{3.18}$$
$$\{$$
$$\text{partition } K \text{ into } K_l = \begin{pmatrix} A_l & B_l \\ B_l^T & C_l \end{pmatrix};$$

    decompose A_l into $A_{FF_v,l}$;

    compute sparse approximation $\tilde{S}_{l,FF_v}$ to $S_l = C_l - B_l^T A_{FF_v,l}^{-1} B_l$;

    if $(l>1)$ $FFDD_v\_setup(\tilde{S}_{l,FF_v},l-1)$;

    else decompose $\tilde{S}_{l,FF_v}$ into S_{l,FF_v} ;

}

$$FFDD_v\_iter(K,l,x,f) \qquad\qquad\qquad (3.19)$$

{

    partition x into $x = (x_\Omega, x_\Gamma)^T$ and f into $f = (f_\Omega, f_\Gamma)^T$;

    compute $\tilde{f}_\Gamma = f_\Gamma - B_l^T A_{FF_v,l}^{-1} f_\Omega$;

    if $(l==0)$

    {

        solve $S_{1,FF_v} x_\Gamma = \tilde{f}_\Gamma$;

    }

    else

    {

        $FFDD_v\_iter(S_{l,FF_v}, l-1, x_\Gamma, \tilde{f}_\Gamma)$;

    }

    solve $A_{FF_v,l} x_\Omega = f_\Omega - B_l x_\Gamma$;

}

All the filtering decompositions on the resp. level and the filtering approximation of the Schur-Complement can be computed in parallel, providing a successively smaller system which at level 0 can then be solved sequentially by one FF step with respect to v.

This new FFDD-algorithm also satisfies the filtering property (2.3) and thus we get in analogy to Remark 2.1

Remark 3.1: *The approximate inverse M generated by the $FFDD_v$ algorithm satisfies (2.4).*

4. An Efficiency Model

It is very important to model the overall efficiency in order to be able to get a close predicition of the real performance of the method. Usually, efficiency is defined by

$$E_{par} = \frac{T_{q=1}}{p \cdot T_{q=p}} \qquad\qquad\qquad (4.1)$$

where T_q means the execution time of one cycle of the same algorithm on q processors.

However, this is only a measure for the load of the computer. Since our aim in parallel processing is to save computing time, we have to take into account the numerical efficiency too. I. e. we have to measure the efficiency of the parallel algorithm used with respect to a sequential one and compare the corresponding convergence rates. Thus, the overall efficiency reads

$$E_{tot} = E_{num} \cdot E_{par} \tag{4.2}$$

with E_{par} from (4.1) and the numerical efficiency

$$E_{num} = \frac{\ln \rho_{par} T_{sk}}{\ln \rho_{sk} T_{par}} \tag{4.3}$$

with the convergence rate per cycle $\rho_{par/sk}$ of the parallel or sequential method resp. and $T_{par/sk}$ the time per cycle of the resp. methods.

We measure the efficiency of the $FFDD_v$ method with respect to the original FF algorithm, which corresponds to $FFDD_v$ for $l_{max} = 0$. The work count for the $FFDD_v$ method depending on l_{max} is given in the following Remark.

Remark 4.1: *The parallel work count of one step of $FFDD_{v\_iter}$ is given by*

$$W(FFDD_{v\_iter}) = 2\,(\,C_{FF} + C_B + 1\,)\,n \sum_{l=0}^{l_{max}} m_l \tag{4.4}$$

with constants C_{FF} and C_B, m_l denoting the number of horizontal gridlines in a strip $\Omega_{h,i}$ on level l (cf. Fig. 1), and n the number of gridpoints on one horizontal gridline. The parallel work count of $FFDD_{v\_setup}$ is given by

$$W(FFDD_{v\_setup}) = 3C_Z\,n \sum_{l=0}^{l_{max}} m_l \tag{4.5}$$

with a constant C_Z.

If we measure the efficiency with constant load („*scaled speedup*"), i.e. we have

$$m \cdot n = C_L \tag{4.6}$$

with a constant C_L,

$$m = m_{lmax}. \tag{4.7}$$

We further require

$$m_l \le m,\ \text{for } i = 0,\ldots,l_{max}\,. \tag{4.8}$$

The corresponding parallel work estimate is given in the following Remark.

Remark 4.2: *Let (4.6), and (4.7) hold. Then the parallel work for the complete FFDD$_v$-method behaves like*

$$W_{tot} = C \ln n.$$ (4.9)

Proof: Directly by inserting (4.6) and (4.8) into (4.4) and (4.5). ○

This estimate shows the main benefit of parallel processing. While the optimal complexity of an algorithm for solving a linear system of equations with N unknowns on a scalar computer is $O(N)$, we are able to obtain $O(\ln N)$ on a parallel machine, if we measuer with constant load (cf. (4.7)).

Unfortunately, a stripe decomposition as used in the present construction will satisfy assumptions (4.6) and (4.7) only for a limited number of processors, as stated in the following Remark.

Remark 4.3: *For a stripe decomposition (cf. Fig. 1) we have*

$$m = \frac{n - p + 1}{2}.$$ (4.10)

Since such a decomposition is reasonable only for m ≥1 we require

$$p \leq \frac{n + 1}{2},$$ (4.11)

otherwise a stripwise decomposition is not reasonable. This is a severe restriction for the applicability of the stripwise domain decomposition for large p.

Incorporating a box splitting of the domain, which readily satisfies assumptions (4.6) and (4.7), it is possible to avoid this restricition. This extension will be the subject of a forthcoming paper.

The above considerations yield the following estimate for the parallel efficiency.

Remark 4.3: *Let (4.6) and (4.7) hold and m = n/p. The parallel efficiency of the FFDD$_v$-algorithm measured versus the scalar FF-method is*

$$E_{par} = \frac{C_{FF}}{(2\,(C_{FF} + C_B + 1) + C_{Comm}p + C_{setup})\ln n}$$ (4.12)

with constants C_{FF} and C_B characterizing the solution process (cf. (4.4) and (4.5)) and constants C_{Comm} and C_{setup} characterizing the data transfer rate in the parallel network and the message setup time of the parallel architecture.

Proof: Inserting (4.4) and (4.5) into definition (4.1) we get

$$E_{par} = \cfrac{C_{FF}\,n}{p\left(2\,(C_{FF}+C_B+1)\,n\sum_{l=0}^{l_{max}} m_l + (C_{Comm}\,n + C_{setup})\,\ln n\right)}.$$

Now using assumptions (4.6), (4.7) and $m = n/p$, we obtain (4.9). $\qquad\qquad$ o

5. Numerical Results

We tested the $FFDD_v$-method on problem (1.1) with $\varphi(x,y) = 1$. To that purpose we implemented the method on a transputer system with 128 T800 processors (Parsytec SC-128) in *ParC* using the *vChannel* routing system (cf. [Ba]). To find out the numerical efficiency we measured the convergence rate of our algorithm for different numbers of processors.

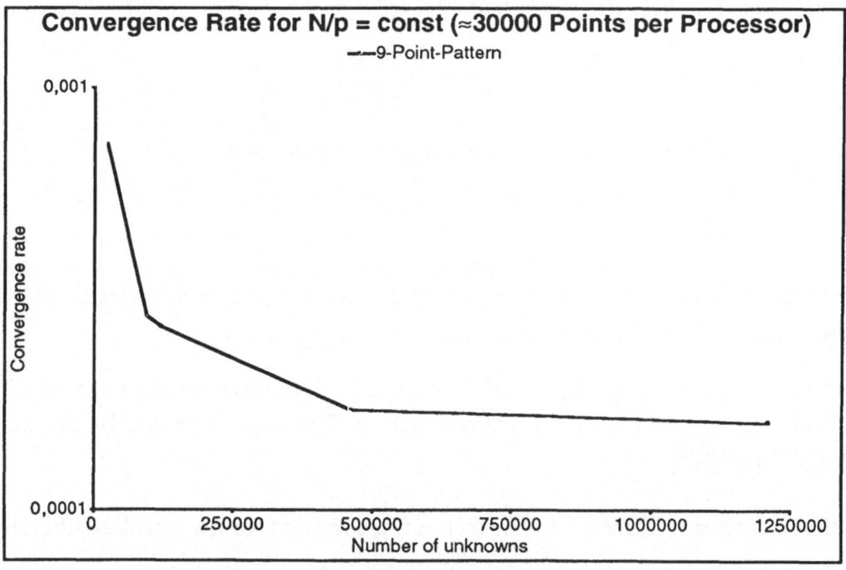

Figure 2: Convergence rate of $FFDD_v$ versus number of unknowns, measured with constant load.

The results are shown in Fig. 2 clearly confirm the optimal numerical efficiency of the parallel algorithm. In this context it has to be mentioned that up to now methods based on a stripwise splitting of the domain did not obtain convergence rates independent of the number of stripes (cf. [Dr]). The improvement of the convergence rate for an increasing number of processors is due to the nonlinear increase of the convergence rate of the scalar FF-method with growing number of unknowns (see [Wi2, Abb. 5.1.2]), which is used as inner solver here.

Then we measured the scaled speedup, i.e. the cpu-time versus the size of the linear system with N/p = const (constant load). The results are given in Fig. 3. They show the factor of about 2 compared with the scalar case ($p=1$), which we have to expect from the construction of the approximate inverse (3.9), since in each outer iteration step we have to apply two inner solves. This basic overhead restricts the overall efficiency to $\leq$ 50%. It is relaxed, however, by the improvement of the convergence rate shown in Fig. 2. Except this limiting factor, the scaling for larger numbers of processors is almost optimal and shows a logarithmic behaviour as predicted in Remark 4.2 (recursive partitioning). In particular this means that for an increase of the number of unknowns from 150.000 to $3 \cdot 10^6$ by a factor of 20, the cpu-time increases from 1.100 to 1.500 sec, corresponding to a factor of 1.36. The resulting scaled speedup is 14.7.

As expected, the method without recursive partitioning (Schur complement sequential) has insufficient scaling properties for larger number of processors ($p>30$). For a small number of processors, the Schur complements are of small size and thus the total cpu time is dominated by the work for the inner solves.

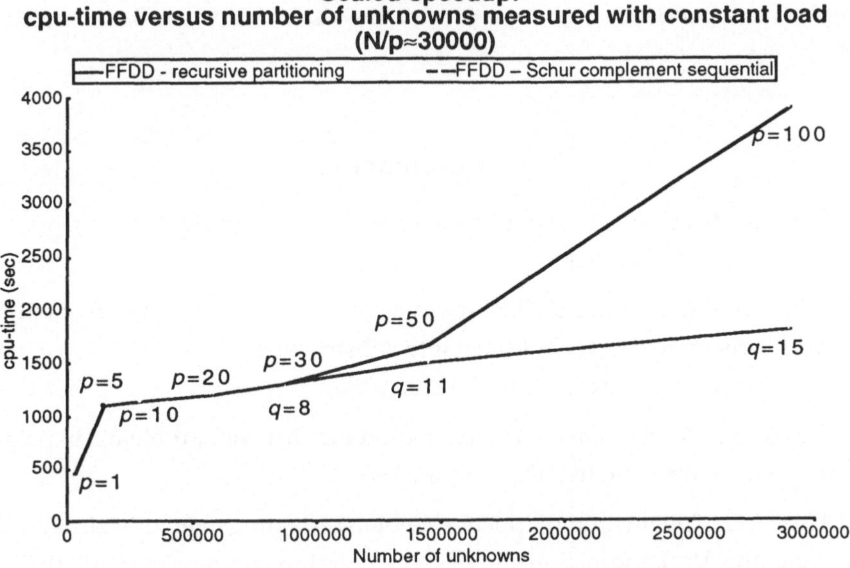

Figure 3: Scaled speedup: cpu-time versus number of unknowns measured with constant load per processor, $N/p = 30.000$. The recursive partitioning (lower curve) used 3 levels and q clusters on level 1, resulting in $m_1 = p/q$.

We finally investigated about the influence of q, the number of clusters on level 1 on the total time needed. The corresponding results are given in Fig. 4. These values show again that a substantial influence shows up only for a large number of processors p, since p is also the size of the Schur complement on level l_{max}.

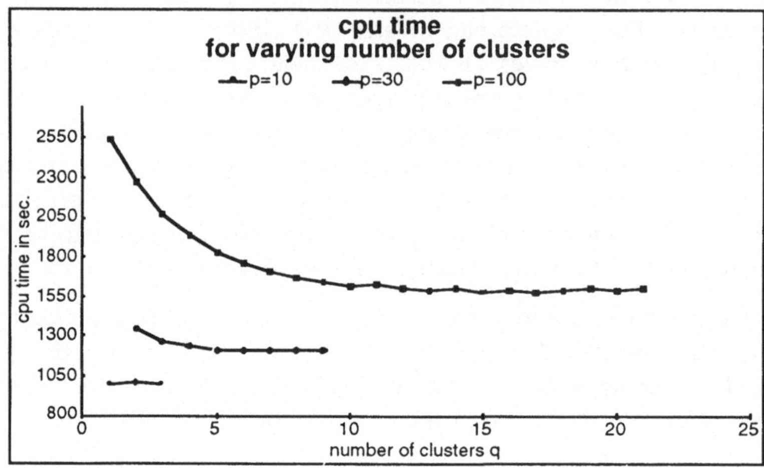

Figure 4: The dependency of the cpu-time on the size of coarse level blocks (3 level method).

References

[Ba] Bastian, P.: *vChannel* – A routing software for Transputer systems.
Preprint, IWR, Heidelberg, 1991

[BW] Björstad, P.E. , Widlund, O.B.: Iterative methods for the solution of elliptic problems on regions partitioned into substructures.
SIAM J. Numer. Anal., 23, 1097-1120, (1986)

[Dr] Dryija, M.: A capacitance matrix method for Dirichlet problems on polygonal regions. Numer. Math., 39 (1), 51-64, 1982

[Ha1] Haase, G.: Die nichtüberlappende Gebietszerlegungsmethode zur Parallelisierung und Vorkonditionierung des CG-Verfahrens. Preprint 92-10, IWR, Universität Heidelberg, 1992

[Ha2] Haase, G., Langer, U., Meyer, A.: The approximate Dirichlet domain decomposition method. Part I: An Algebraic Approach.
Computing 47, 137-151 (1991)

[Ha3] Haase, G., Langer, U., Meyer, A.: The approximate Dirichlet domain decomposition method. Part II: Applications to 2nd-order elliptic b.v.p.s. Computing 47, 153-167 (1991)

[Wi1] Wittum, G.: An ILU-based smoothing correction scheme. inHackbusch,W.(ed.): Parallel solvers. Proceedings of the sixth GAMM-Seminar, Kiel, Jan. 25 to 27, 1990. Notes on numerical fluid mechanics, Vieweg, Braunschweig i. Vb.

[Wi2] Wittum, G.: Frequenzfilternde Zerlegungen – Ein Beitrag zur schnellen Lösung großer Gleichungssysteme. Teubner Skripten zur Numerik Band 1, Teubner, Stuttgart, 1992

THE ROLE OF INCOMPLETE LU-FACTORIZATION IN MULTIGRID METHODS

P. Wesseling
Delft University of Technology,
Department of Technical Mathematics and Informatics,
P.O. Box 5031, 2600 GA Delft,
The Netherlands

INTRODUCTION

Since their introduction as preconditioners for conjugate gradient methods (ICCG method, [37]) and the widely appreciated demonstration of the efficiency of these preconditioners in [27], incomplete LU (ILU) factorizations have found widespread use in the context of conjugate gradient and related methods. Not long after its introduction, ILU was also applied as smoother in a multigrid method and comparison was made with a conjugate gradient method, preconditioned with the same ILU factorization, for a Navier-Stokes application ([55]). More extensive comparisons of conjugate gradients and multigrid as linear systems solvers are reported in [45] and [12]. Viewed as linear systems solvers, conjugate gradients and multigrid can be regarded as two alternate ways to accelerate convergence of basic iterative methods. Generally speaking, for medium-sized linear problems conjugate gradients and multigrid are about equally effective, but for sufficiently large problems multigrid is faster, because of its linear computational complexity. Unlike multigrid, conjugate gradient methods are limited to linear problems. On the other hand, conjugate gradient methods are much easier to program, especially when the computational grid is unstructured.

Here the basics of the use of ILU factorizations in the multigrid context are reviewed. To assist the uninitiated reader of this volume, we start with a brief introduction to multigrid methods.

MULTIGRID METHODS

Extensive introduction to multigrid methods is provided in [19] and [51], to which we refer for further details. Here we confine ourselves to linear problems in two dimensions. The system to be solved is assumed to originate in the discretization of a partial differential equation on a uniform computational grid G^K, defined by

$$G^K = \{x \in I\!R^2 : x = jh, \ j = (j_1, j_2), \ j_\alpha = 0, 1, 2, ..., 2^K, \ h = 1/n\}. \tag{1}$$

A sequence of coarser grids G^k, $k = K, K-1, ..., 1$ is defined by (1) with K replaced by k (*vertex-centered coarsening*). The linear algebraie system to be solved is denoted by

$$A^K u^K = b^K. \tag{2}$$

Let $U^k : G^k \to I\!R$ be the set of grid functions on G^k. *Prolongation* operators $P^k : U^{k-1} \to U^k$ and *restriction* operators $R^k : U^{k+1} \to U^k$ are defined. Prolongation may be defined by interpolation, a possibility for restriction is to take the adjoint of the prolongation:

$$R^{k-1} = (P^k)^*. \tag{3}$$

The transfer operators have to satisfy the following accuracy rule on all grids:

Here m_P is defined as one plus the highest degree of polynomials that are interpolated exactly by P^k, and m_R is defined in the same way, using $(R^k)^*$; $2m$ is the order of the differential equation that is approximated by (2).

The main ingredients of multigrid methods are *smoothing* and *coarse grid approximation*. For smoothing we take a basic iterative method, defined as follows. Let A^k be split as $A^k = M^k - N^k$, then the method is defined as

$$M^k(u^k)^{m+1} = b^k + N^k(u^k)^m, \tag{5}$$

where m is an iteration index. The basic principle of multigrid methods is that the error is made smooth by just a few iterations of (5) (therefore this process is called smoothing), whereas the smooth part of the error is approximated on coarser grids. This idea leads to the

Two-grid algorithm
begin
(1) $S(u, v, b, K)$ /* do (5) on G^K with u the new and v the old approximation */
(2) $r^K = b^K - A^K u^K$
(3) Solve $A^{K-1} u^{K-1} = R^{K-1} r^K$
(4) $u^k = u^K + P^K u^{K-1}$
(5) $S(u, u, b, K)$
end

This defines one two-grid iteration. The *coarse grid matrix* A^{K-1} may be obtained by discretizing the partial differential equation on G^{K-1}, or by

$$A^{K-1} = R^{K-1} A^K P^K. \tag{6}$$

A multigrid method is obtained if coarse grid solution (statement (3)) is replaced by γ multigrid iterations. This gives us the following recursive subroutine:

subroutine $MG(u, v, b, K)$
begin
    if (k **eq** 1) **then**
        Solve $A^1 u^1 = b^1$
    else
        $S(u, v, b, k)$
        $b^{k-1} = R^{k-1}(b^k - A^k u^k)$
        $v^{k-1} = 0$
        for $i = 1$ **step** 1 **until** γ **do**
            $MG(u, v, b, k - 1)$
        done
        $u^k = u^k + P^k u^{k-1}$
        $S(u, u, b, k)$
    end if
end of MG

A structure diagram for a non-recursive version of this algorithm may be found in [51]. The order in which the grids are visited is called the *multigrid cycle* or *schedule*. With $\gamma = 1$ one obtains the *V-cycle*, with $\gamma = 2$ the *W-cycle*. The simplest cycle is the *sawtooth cycle*, obtained from the V-cycle by deleting one of the smoothing operations, resulting in (recursiveness evaporates for $\gamma = 1$):

Sawtooth cycle
begin

$$b^{K-1} = R^{K-1}(b^K - A^K u^K)$$

for $k = K - 2$ **step** -1 **until** 1 **do**

$$b^k = R^k b^{k+1}$$

done

Solve $A^1 u^1 = b^1$

for $k = 2$ **step** 1 **until** $K - 1$ **do**

$$u^k = P^k u^{k-1}$$

$$S(u, u, b, k)$$

done

$$u^K = u^K + P^K u^{K-1}$$

$$S(u, u, b, K)$$

end

The main property of multigrid is

$$W = O(N), \tag{7}$$

where W is the amount of computational work and N is the number of unknowns. Property (7) can be realized for many cases. For proofs see [4] (an early proof) or [19] (survey of the theory). An early paper where (7) is demonstrated convincingly in practice is [7].

APPROXIMATE INVERSE

Jacobi ([14], [4]) and Gauss-Seidel ([8], [7], [17]) iteration were the first smoothing methods used in multigrid, and continue to be used today. Another early smoother that is now only of historical interest is the *approximate inverse* ([15]), which we will now describe. But first we will introduce *stencil notation*. The grid superscript k can now be omitted. A linear operator $A : U \to U$ can be represented as

$$Au_i = \sum_j A(i, j) u_{ij}, \tag{8}$$

where i, j are grid coordinates according to (1). For given i, the set of values $A(i, j)$ is called the stencil $[A]_i$ of A at i, with the familiar representation

$$[A]_i = \begin{bmatrix} \cdots & A(i,(0,1)) & \cdots \\ A(i,(-1,0)) & A(i,(0,0)) & A(i,(1,0)) \\ \cdots & A(i,(0,-1)) & \cdots \end{bmatrix}. \tag{9}$$

In [15] the approximate inverse B of A is defined as follows. Let the structure (non-zero elements) of $[A]$ (for all i) be given by

$$[A] = \begin{bmatrix} * & * & * \\ * & * & * \\ * & * & * \end{bmatrix}. \tag{10}$$

The structure of $[B]$ is chosen to be the same, and B is defined by

$$\sum_{|j| \leq 1} A(i + j, k - j) B(i, j) = \delta_{k,0}, \ \forall |k| \leq 1, \tag{11}$$

where $|j| = \max\{|j_1|, |j_2|\}$ and $\delta_{k,0}$ is the Kronecker delta. This implies that

$$[BA] = \begin{bmatrix} \dots & * & * & * & * & * & \dots \\ \dots & * & 0 & 0 & 0 & * & \dots \\ \dots & * & 0 & 1 & 0 & * & \dots \\ \dots & * & 0 & 0 & 0 & * & \dots \\ \dots & * & * & * & * & * & \dots \end{bmatrix}. \tag{12}$$

In other words, on the non-zero pattern of A, BA cannot be distinguished from the identy. According to (11), B is computed row by row, solving a 9×9 system. Smoothing takes place according to

$$u^{(m+1)} = u^{(m)} + B(b - Au^{(m)}). \tag{13}$$

Frederickson goes on to develop a multigrid method, using bilinear interpolation for P^k and choosing $R^{k-1} = (P^k)^*$, defining $A^k, k < K$ by recursive application of (6), and using the sawtooth cycle defined in the preceding section. The resulting multigrid algorithm shows an asymptotic reduction factor of 0.45 for Poisson's equation. Replacing (13) by an ILU-iteration (defined in the next section) one finds an asymptotic reduction factor of 0.07, for a comparable amount of work. Frederickson's method is generalized to the vorticity-stream function formulation of the Navier-Stokes equations in [52], with asymptotic reduction factors between 0.3 and 0.5. Again replacing (13) by ILU, the asymptotic reduction factors obtained are about 0.1 ([55]).

Because of its greater efficiency, ILU has superseded the approximate inverse. However, the basic idea behind the two methods shows similarity. Switching to standard matrix notation, let G be an index set of matrix elements. Then the approximate inverse B is defined by

$$(BA)_{ij} = I_{ij}, \quad (i,j) \in G, \tag{14}$$

for a suitable choice of G, with I the identity matrix. *Incomplete LU factorization* is defined by, for example (more details in the following section)

$$(LU)_{ij}) = A_{ij}, \quad (i,j) \in G. \tag{15}$$

The efficiency of a smoothing method for a particular problem may be measured by means of Fourier analysis, to be descussed later. Fourier smoothing analysis has not been carried out for approximate inverse smoothing.

INCOMPLETE FACTORIZATION

Incomplete LU factorization is described in [37] and [36]. In (5) we choose (deleting the grid superscript k)

$$M = LD^{-1}U, \tag{16}$$

with $diag(L) = diag(U) = D$, D a diagonal matrix, and L and U lower and upper triangular. L, D and U are determined as follows. A graph G of the incomplete decomposition is defined, consisting of the index set (i,j) for which L_{ij}, D_{ij} and U_{ij} are allowed to be non-zero. We require

$$(LD^{-1}U)_{ij} = A_{ij}, \quad \forall (i,j) \in G. \tag{17}$$

On an $I \times J$ grid one may choose for example (assuming lexicographic ordering)

$$G = \{(i,j) \in [1,I] \times [1,J] : j = i, i \pm 1, i \pm I \mp 1, i \pm I\}, \tag{18}$$

resulting in 7-point ILU. Equation (17) leads to recursion formulas from which the elements of L, D and U are easily determined; see for example [37], [36], [45] or [51]. *Modified incomplete factorization* is obtained if D as found from (17) is changed to $D + \sigma\tilde{D}$, with σ a parameter and $\tilde{D}$ a diagonal matrix with elements

$$\tilde{D}_{ii} = \sum_{j \neq i} |N_{ij}|, \tag{19}$$

where $N = LD^{-1}U - A$. The optimal value of σ is problem dependent. If one values robustness there should not be a free parameter in the smoother. Therefore we will not look for the optimal σ but always use either $\sigma = 0$ or $\sigma = 1/2$.

In [37] it is shown that if A is an M-matrix, then L, D and U as defined by (17) exist and can be stably computed. In [56], [57] it is shown that with suitable σ modified ILU has the *smoothing property* as defined in [19].

Alternating ILU consists of one ILU iteration of the type just described followed by a second ILU iteration based on a backward ordering of the grid points, given by (the pair (i, j) are grid point coordinates)

$$k = IJ + 1 - j - (i - 1)I, \tag{20}$$

resulting in $\tilde{M} = \tilde{L}\tilde{D}^{-1}\tilde{U}$. The stencils of $L, U, \tilde{L}$ and $\tilde{U}$ are given by (7-point incomplete factorization)

$$[L]_i = \begin{bmatrix} 0 & 0 & 0 \\ \gamma_i & \delta_i & 0 \\ 0 & \alpha_i & \beta_i \end{bmatrix}, \quad [U]_i = \begin{bmatrix} \xi_i & \eta_i & 0 \\ 0 & \delta_i & \mu_i \\ 0 & 0 & 0 \end{bmatrix},$$

$$[\tilde{L}]_i = \begin{bmatrix} 0 & \tilde{\gamma}_i & 0 \\ 0 & \tilde{\delta}_i & \tilde{\alpha}_i \\ 0 & 0 & \tilde{\beta}_i \end{bmatrix}, \quad [\tilde{U}]_i = \begin{bmatrix} \tilde{\xi}_i & 0 & 0 \\ \tilde{\eta}_i & \tilde{\delta}_i & 0 \\ 0 & \tilde{\mu}_i & 0 \end{bmatrix}, \tag{21}$$

where now $i = (i_1, i_2)$ represents grid point coordinates.

Another type of incomplete factorization is *incomplete block LU-factorization* (IBLU). Let the stencil of A be of type (10). Then A has the following structure:

$$A = \begin{pmatrix} B_1 & U_1 & & & \\ L_2 & B_2 & U_2 & & \\ & \ddots & \ddots & \ddots & \\ & & \ddots & \ddots & U_{J-1} \\ & & & L_J & B_J \end{pmatrix}, \tag{22}$$

with L_J, B_j and U_j $\ I \times I$ tridiagonal matrices. Define

$$D_1 = B_1, \quad D_j = B_j - L_j D_{j-1}^{-1} U_j, \quad j = 2, 3, \ldots J. \tag{23}$$

Provided D_j^{-1} exist we find that

$$A = (L + D)D^{-1}(D + U), \tag{24}$$

with

$$\begin{pmatrix} 0 & & & \\ L_2 & 0 & & \\ & \ddots & \ddots & \\ & & L_j & 0 \end{pmatrix}, \quad U = \begin{pmatrix} 0 & U_1 & & \\ & 0 & \ddots & \\ & & \ddots & U_{J-1} \\ & & & \end{pmatrix}. \tag{25}$$

From this complete factorization (called a block factorization or line factorization because L_j, D_j and U_j correspond to (in our case horizontal) lines in the grid) an incomplete factorization is obtained by replacing the full matrices D_j by sparse approximations $\tilde{D}_j$, for example by approximating (21) as folows:

$$\tilde{D}_1 = B_1, \quad \tilde{D}_j = B_j - tridiag(L_j \tilde{D}_{j-1}^{-1} U_j), \quad j = 2, 3, ..., J. \tag{26}$$

An iterative method is obtained by choosing

$$M = (L + \tilde{D})\tilde{D}^{-1}(\tilde{D} + U) \tag{27}$$

in (5). For algorithms to compute $\tilde{D}_j$, see [2], [10], [3], [41], [44] and [51]. The first four publications give existence proofs for $\tilde{D}_j$ if A is an M-matrix.

ROBUSTNESS AND EFFICIENCY

In order to investigate and compare efficiency and robustness of solution methods for discretizations of partial differential equations, the following two test problems are useful:

$$-(\varepsilon c^2 + s^2)\frac{\partial^2 u}{\partial x^2} - 2(\varepsilon - 1)cs\frac{\partial^2 u}{\partial x \partial y} - (\varepsilon s^2 + c^2)\frac{\partial^2 u}{\partial y^2} = 0 \tag{28}$$

$$-\varepsilon(\frac{\partial^2 u}{\partial x^2} + \frac{\partial^2 u}{\partial y^2}) + c\frac{\partial u}{\partial x} + s\frac{\partial u}{\partial y} = 0, \tag{29}$$

with $c = \cos\beta$, $s = \sin\beta$. There are two constant parameters to be varied: $\varepsilon > 0$ and $0 \le \beta < 2\pi$. The rotated anisotropic diffusion equation (28) models anisotropic diffusion and/or high mesh aspect ratios, and contains a mixed derivative. The convection-diffusion equation (29) models non-self-adjoint and almost hyperbolic ($\varepsilon \ll 1$) problems. We call a method robust if its efficiency is uniform in ε and β. Equation (28) is discretized by central differences. The mixed derivative is discretized with the following stencil:

$$\frac{1}{h^2}\begin{bmatrix} -1 & 1 & 0 \\ 1 & -2 & 1 \\ 0 & 1 & -1 \end{bmatrix}. \tag{30}$$

Due to the presence of positive off-diagonal elements the resulting matrix is not an M-matrix if $(\varepsilon - 1)cs < 0$. For a discretization that gives an M-matrix, see [51]. Equation (29) is discretized with upwind differences, so that we obtain an M-matrix.

The efficiency of a smoothing method can be measured with the *Fourier smoothing factor ρ*, introduced in [7]. For an introduction to Fourier smoothing analysis and a survey of the literature, see [51]. The smoothing factor ρ measures the reduction of non-smooth Fourier components of the error in a smoothing step. Fourier smoothing analysis assumes periodic boundary conditions. Dirichlet boundary conditions may be taken into account in a heuristic way, resulting in a smoothing factor $\rho_D \le \rho$, with $\lim_{h \to 0} \rho_D = \rho$ with ε fixed. For practical values of h the difference between ρ_D and ρ may be appreciable, in which case ρ_D predicts more accurately the multigrid convergence behaviour than ρ. For details, see [51].

We now present ρ and ρ_D for a number of cases. Robustness and efficiency are investigated by varying ε in the set $\{1, 10^{-1}, 10^{-2}, 10^{-3}, 10^{-5}\}$ and β in the set $\{\beta = k\pi/12 : k = 0, 1, 2, ..., 23\}$. The grid is assumed to be uniform and to consist of $n \times n$ points with $n = 64$. Unless stated otherwise, increasing n or diminishing ε further has no influence to speak of. Table 1 gives results for 7-point ILU with $\sigma = 0$ (unmodified) and $\sigma = 1/2$ (modified) for (29). Only the

Table 1   Fourier smoothing factors for the convection-diffusion equation; 7-point ILU.

		$\sigma = 0$		$\sigma = 1/2$	
ε	β	ρ	ρ_D	ρ	ρ_D
10^{-5}	165^0	0.58	0.54	0.47	0.47

For $\beta = 0^0, 180^0$ and $\varepsilon \ll 1$ we have $\rho \cong 0$. However, for $\varepsilon \ll 1$ there is a small neighbourhood of $\beta = 0^0$ and $\beta = 180^0$ where ρ and ρ_D get close to 1 with $\sigma = 0$; this situation is explored in detail in [51]. For example, for $\beta = 174^0$ we find $\rho = 0.82$. Since this occurs only for a very limited range of β and since for many values of β we have ρ small (< 0.1), in practice 7-point ILU with $\sigma = 0$ is still found to be a good smoother. With $\sigma = 1/2$ this difficulty does not occur, and we have a robust smoother for (29). Table 2 gives results for alternating ILU.

Table 2   Fourier smoothing factors for the convection-diffusion equation; alternating 7-point ILU.

		$\sigma = 0$	$\sigma = 1/2$
ε	β	ρ, ρ_D	ρ, ρ_D
10^{-5}	105^0	0.086	0.036

Clearly, alternating 7-point ILU is robust and very efficient for (29).

Results for the rotated anisotropic diffusion equation (28) are presented in Table 3.

Table 3   Fourier smoothing factors for the rotated anisotropic diffusion equation; 7-point ILU.

		$\sigma = 0$		$\sigma = 1/2$	
ε	β	ρ	ρ_D	ρ	ρ_D
10^{-5}	75^0	1.92	1.66	0.68	0.68
10^{-5}	90^0	0.98	0.002	0.33	0.001

The smoother is not robust for $\sigma = 0$. But also for $\sigma = 1/2$ this smoother is not very effective. For example, with finer sampling of β around 75^0 one finds $\rho_D = 0.73$ for $\beta = 85^0$. If there is no mixed derivative ($\beta = 0^0$ or 90^0), then this is a good smoother. Note the large difference between ρ and ρ_D. It can be shown ([51]) that for $0 \leq \sigma \leq 1/2$ and $\varepsilon \ll 1$ ρ_D satisfies

$$\rho_D \cong |(\sigma - 1 + 2\varphi^2)/\{\delta^2(2 + \varphi^2/2\varepsilon) + \sigma - 1\}| \tag{31}$$

where $\delta \cong 1 + \sqrt{2\varepsilon(1 + \sigma)}$ and $\varphi = 2\pi/n$.
Table 4 gives results for alternating ILU.

Table 4   Fourier smoothing factors for the rotated anisotropic diffusion equation; alternating 7-point ILU.

		$\sigma = 0$	$\sigma = 1/2$
ε	β	ρ, ρ_D	ρ, ρ_D
10^{-4}	45^0	0.94	0.11

We see that modified ($\sigma = 1/2$) alternating 7-point ILU is very efficient smoother for all cases.

The use of alternating ILU has been proposed in [39]. Modification has been analyzed and tested in [23], [39], [32], [56], [57].

About the IBLU smoother we can be brief. For (28) the worst case encountered is $\varepsilon = 10^{-5}$, $\beta = 45^0$, giving $\rho = \rho_D \cong 0.17$. This is a rough approximation, for reasons explained in [51]. The worst cases for (29) are $\varepsilon = 10^{-5}$, $\beta = 0^0$, $\rho = 0.2$ and $\varepsilon = 10^{-3}$, $\beta = 0^0$, $\rho_D = 0.12$. This is a very robust and efficient smoother.

Not much work has yet been done on incomplete factorization smoothing in three dimensions. In [30] Fourier smoothing analysis is carried out for a three-dimensional version of IBLU. For the following equation

$$\frac{\partial^2 u}{\partial x^2} + \varepsilon \frac{\partial^2 u}{\partial y^2} + \varepsilon \frac{\partial^2 u}{\partial z^2} = 0 \tag{32}$$

or equations obtained by permuting ε, this smoother is efficient for $0 < \varepsilon \leq 1$, but the smoothing factor $\rho \uparrow 1$ as $\varepsilon \to \infty$. An extensive study of incomplete factorization smoothing in two and three dimensions is given in [47], [31] and [32]. A completely robust three-dimensional smoother is not found, unless one allows smoothers involving accurate solution in planes (alternating plane smoothing).

MULTIGRID VERSUS CONJUGATE GRADIENTS

Here we will consider multigrid methods only as solvers for linear algebraic systems arising from nine- (or fewer) point discretizations of partial differential equations. In this context, like conjugate gradients, multigrid can be viewed as an acceleration technique for basic iterative methods, and here the two will be compared. Of course, the scope of multigrid is much wider. For the nonlinear multigrid algorithm, or the application of multigrid to other types of problems, see for example [19], [9] or [51].

If $A^k, k < K$ is generated by (6) a linear multigrid code can be presented to the user like any other linear algebraic systems solver, requiring only input of the matrix and the right-hand-side on the finest grid. The resulting code will be of *autonomous* or *black box* type, i.e. it does not require any action from the user apart from problem specification. Examples of such codes, using incomplete factorization, are MGD1 and MGD5 ([53], [24], [26], [25], [45], [44]), which are available in the NAG library. The multigrid algorithm is the sawtooth cycle as defined in the proceding section. MGD1 uses ILU, MGD5 uses IBLU for smoothing. Recent updates that are efficient on vector computers and a version that handles discontinuous coefficients (MGCS, [11]) are available from P.M. de Zeeuw (Center for Mathematics and Informatics, P.O. Box 4079, 1009 AB Amsterdam, telefax +3120-5924199, email pauldz@cwi.nl).

The conjugate gradient algorithm for symmetric positive definite systems is well-known; for an introduction see [22], [16] or [18]. For nonsymmetric cases generalizations have been proposed, such as ORTHOMIN ([48]), GMRES ([42]) and CGS ([43]). Comparative numerical experiments with conjugate gradients and multigrid are presented in [6], [5], [55] and [45]. One can also combine conjugate gradients and multigrid, using multigrid as preconditioner for conjugate gradients, so that multigrid is accelerated by conjugate gradients. The amount of additional computing involved is small. The rate of convergence is never impaired, and may improve significantly when multigrid is slow, as may happen when the smoother is not efficient or coarse grid approximation is not accurate. Combinations of conjugate gradients and multigrid are described in [28], [6], [5] and [45].

Figure 1 (from [45], with permission) shows a typical result for a test problem proposed by Kershaw ([27]). Two basic iterative methods (7-point ILU and IBLU) are accelerated by conjugate gradients (CG) or multigrid (MG); MGCG is the combination of MG and CG. The dimension of the grid is 51 × 51. In order to allow a sufficient number of coarse grids by mesh-doubling the grid was increased to 57 × 57 in the multigrid case by adding artificial equations (padding); this gives a coarsest grid of 8 × 8. Padding entails an additional burden of computing work, which could be diminished by tampering with the code, but this would go against the black box philosophy, so this was not done. If this work were to be done again we would include the alternating and modified ($\sigma \neq 0$) ILU versions. When increasing the problem size, CG will start to lag behind MG, but we see that for medium sized problems CG and MG are about equally efficient. For modest precision MG is faster, especially if *nested iteration* (see [19] or [51]; not discussed here) is applied.

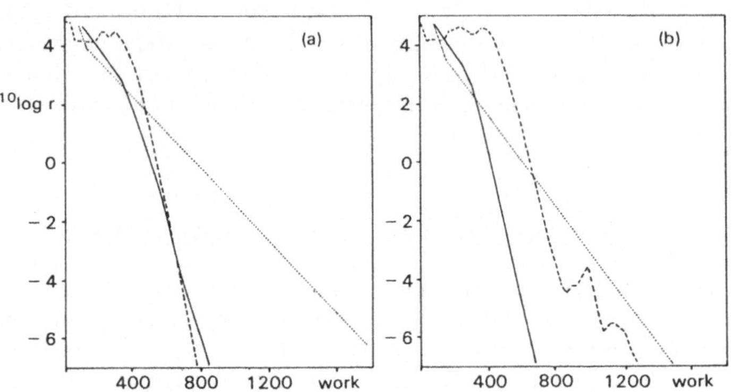

Figure 1   Convergence history for Kershaw's problem; (a): IBLU, (b): ILU; —: MGCG, - - -: CG, ... : MG

The test problem of Kershaw involves strongly discontinous coefficients. This makes it necessary to use matrix-dependent prolongation and restriction operators ([29], [1], [11]) in multigrid (unless cell-centered multigrid is used (not considered here), cf. [49], [50], [54], [33], [46]).

CONCLUDING REMARKS

A review of the use of incomplete factorizations as smoothers in multigrid methods has been given, and comparison has been made with methods of conjugate gradient type. For the class of problems discussed here no rate convergence theory predicting rates of convergence observed in practice is available. Therefore, numerical experiments are necessary for comparison and validation purposes. Some suitable test problems have been presented. In two dimensions efficient and robust incomplete factorization smoothers and preconditioners are available, especially IBLU and alternating modified ILU. With such smoothers dependable black box linear multigrid subroutines, such as MGD1, MGD5 and MGCS can be constructed. For medium sized problems such codes are about as efficient as methods of conjugate gradient type, which are easier to program and easier to apply to general sparse matrices. But for large problems multigrid is faster.

References

[1] R.E. Alcouffe, A. Brandt, J.E. Dendy Jr., and J.W. Painter. The multigrid method for diffusion equations with strongly discontinuous coefficients. *SIAM J. Sci. Stat. Comp. 2*,

[2] O. Axelsson, S. Brinkkemper, and V.P. Il'in. On some versions of incomplete block matrix factorization iterative methods. *Lin. Algebra Appl. 59*, 3–15, 1984.

[3] O. Axelsson and B. Polman. On approximate factorization methods for block matrices suitable for vector and parallel processors. *Lin. Alg. Appl. 77*, 3–26, 1986.

[4] N.S. Bachvalov. On the convergence of a relaxation method with natural constraints on the elliptic operator. *USSR Comp. Math. and Math. Phys 6*, 101–135, 1966.

[5] A. Behie and P.A. Forsyth, Jr. Comparison of fast iterative methods for symmetric systems. *IMA J. Numer. Anal.*, 3:41–63, 1983.

[6] A. Behie and P.A. Forsyth, Jr. Multi-grid solution of the pressure equation in reservoir simulation. In *Proc. Sixth Annual Meeting of Reservoir Simulation*, Society of Petroleum engineers, New Orleans, 1982.

[7] A. Brandt. Multi-level adaptive solutions to boundary value problems. *Math. Comp.*, 31:333–390, 1977.

[8] A. Brandt. Multi-level adaptive technique (MLAT) for fast numerical solution to boundary value problems. In H. Cabannes and R. Temam, editors, *Proc. Third Int. Conf. on Numerical Methods in Fluid Mechanics, Vol. 1*, pages 82–89, Springer-Verlag, Berlin, 1973. Lecture Notes in Physics 18.

[9] A. Brandt. The Weizmann Institute research in multilevel computation: 1988 report. In *J. Mandel et al. (1989)*, pages 13–53.

[10] P. Concus, G.H. Golub, and G. Meurant. Block preconditioning for the conjugate gradient method. *SIAM J. Sci. Stat. Comp. 6*, 220–252, 1985.

[11] P.M. de Zeeuw. Matrix-dependent prolongations and restrictions in a blackbox multigrid solver. *J. of Comp. and Appl. Math.*, 3:1–27, 1990.

[12] J.E. Dendy, Jr. and J.M. Hyman. Multi-grid and ICCG for problems with interfaces. In M.H. Schultz, editor, *Elliptic problem solvers*, pages 247–253, Academic Press, New York, 1981.

[13] B. Enquist and T. Smedsaas, editors. *PDE software: Modules interfaces and systems*, North-Holland, Amsterdam, 1984. Procs. IFIP WG 2.5 Working Conference.

[14] R.P. Fedorenko. The speed of convergence of one iterative process. *USSR comp. Math. and Math. Phys.*, 4 no. 3:227–235, 1964.

[15] P.O. Frederickson. *Fast approximate inversion of large elliptic systems*. Report 7–75, Lakehead University, Thunderbay, Canada, 1975.

[16] G.H. Golub and C.F. Van Loan. *Matrix computations*. Johns Hopkins University Press, Baltimore, 1989.

[17] W. Hackbusch. *Ein iteratives Verfahren zur schnellen Auflösung elliptischer Randwertprobleme*. Report 76–12, Universität Köln, 1976.

[18] W. Hackbusch. *Iterative Lösung grosser schwachbesetzter Gleichungssysteme*. Teubner, Stuttgard, 1991.

[19] W. Hackbusch. *Multi-grid methods and applications*. Springer-Verlag, Berlin, 1985.

[20] W. Hackbusch, editor. *Robust Multi-Grid Methods*, Vieweg, Braunschweig, 1989. Proc. 4th GAMM-Seminar, Kiel, 1988. Notes on Numerical Fluid Mechanics 23.

[21] W. Hackbusch and U. Trottenberg, editors. *Multigrid methods*, Springer-Verlag, Berlin, 1982. Lecture Notes in Mathematics (960).

[22] L.A. Hageman and D.M. Young. *Applied iterative methods*. Academic Press, New York, 1981.

[23] P.W. Hemker. The incomplete LU-decomposition as a relaxation method in multi-grid algorithms. In J.H. Miller, editor, *Boundary and Interior Layers - Computational and Asymptotic Methods*, pages 306–311, Boole Press, Dublin, 1980.

[24] P.W. Hemker, R. Kettler, P. Wesseling, and P.M. De Zeeuw. Multigrid methods: development of fast solvers. *Appl. Math. Comp.*, 13:311–326, 1983.

[25] P.W. Hemker and P.M. de Zeeuw. Some implementations of multigrid linear systems solvers. In *Paddon and Holstein (1985)*, pages 85–116.

[26] P.W. Hemker, P. Wesseling, and P.M. de Zeeuw. A portable vector-code for autonomous multigrid modules. In *Enquist and Smedsaas (1984)*, pages 29–40.

[27] D. S. Kershaw. The incomplete Choleski-conjugate gradient method for the iterative solution of systems of linear equations. *J. Comp. Phys.*, 26:43–65, 1978.

[28] R. Kettler. Analysis and comparison of relaxation schemes in robust multigrid and conjugate gradient methods. In *Hackbusch and Trottenberg (1982)*, pages 502–534.

[29] R. Kettler and J.A. Meijerink. *A Multigrid Method and a Combined Multigrid-Conjugate Gradient Method for Elliptic Problems with Strongly Discontinuous Coefficients in General Domains*. Technical Report Shell Publ. 604, KSEPL, Rijswijk, The Netherlands, 1981.

[30] R. Kettler and P. Wesseling. Aspects of multigrid methods for problems in three dimensions. *Appl. Math. Comp.*, 19:159–168, 1986.

[31] M. Khalil. *Analysis of Linear Multigrid Methods for Elliptic Differential Equations with Discontinuous and Anisotropic Coefficients*. PhD thesis, Delft University of Technology, Delft, The Netherlands, 1989.

[32] M. Khalil. Local mode smoothing analysis of various incomplete factorization iterative methods. In *Hackbusch (1989)*, pages 155–164.

[33] M. Khalil and P.Wesseling. Vertex-centered and cell-centered multigrid for interface problems. *J. Comp. Phys.*, 98:1–10, 1992.

[34] J. Mandel, S.F. McCormick, J.E. Dendy, Jr., C. Farhat, G. Lonsdale, S.V. Parter, J.W. Ruge, and K. Stüben, editors. *Proceedings of the Fourth Copper Mountain Conference on Multigrid Methods*, SIAM, Philadelphia, 1989.

[35] S.F. McCormick, editor. *Multigrid methods*, Marcel Dekker Inc., New. York, 1988. Lecture Notes in Pure and Applied Mathematics 110.

[36] J.A. Meijerink and H.A. Van der Vorst. Guidelines for the usage of incomplete decompositions in solving sets of linear equations as they occur in practical problems. *J. Comput. Phys.*, 44:134–155, 1981.

[37] J.A. Meijerink and H.A. Van der Vorst. An iterative solution method for linear systems of

[38] K.W. Morton and M.J. Baines, editors. *Numerical Methods for Fluid Dynamics II*, Clarendon Press, Oxford, 1986.

[39] K.-D. Oertel and K. Stüben. Multigrid with ILU-smoothing: systematic tests and improvements. In *Hackbusch(1989)*, pages 188–199.

[40] D.J. Paddon and H. Holstein, editors. *Multigrid Methods for Integral and Differential Equations*, Clarendon Press, Oxford, 1985.

[41] B. Polman. Incomplete blockwise factorizations of (block) H-matrices. *Lin. Alg. Appl.*, 90:119–132, 1987.

[42] Y. Saad and M.H. Schultz. GMRES: a generalized minimal residual algorithm for solving non-symmetric linear systems. *SIAM J. Sci. Stat. Comp.*, 7:856–869, 1986.

[43] P. Sonneveld. CGS, a fast Lanczos-type solver for nonsymmetric linear systems. *SIAM J. Sci. Stat. Comput.*, 10:36–52, 1989.

[44] P. Sonneveld, P. Wesseling, and P.M. de Zeeuw. Multigrid and conjugate gradient acceleration of basic iterative methods. In *Morton and Baines (1986)*, pages 347–368, 1986.

[45] P. Sonneveld, P. Wesseling, and P.M. de Zeeuw. Multigrid and conjugate gradient methods as convergence acceleration techniques. In *Paddon and Holstein (1985)*, pages 117–168, 1985.

[46] R. Teigland and G.E. Fladmark. Cell-centered multigrid methods in porous media flow. In W. Hackbusch and U. Trottenberg, editors, *Multigrid methods III*, page , 1991.

[47] C.-A. Thole and U. Trottenberg. Basic smoothing procedures for the multigrid treatment of elliptic 3D-operators. *Appl. Math. Comp.*, 19:333–345, 1986.

[48] P.K.W. Vinsome. *ORTHOMIN, an iterative method for solving sparse sets of simultaneous linear equations*. Society of Petroleum Engineers, 1976. paper SPE 5729.

[49] P. Wesseling. Cell-centered multigrid for interface problems. *J. Comp. Phys.*, 79:85–91, 1988.

[50] P. Wesseling. Cell-centered multigrid for interface problems. In *In McCormick (1988)*, pages 631–641, 1988.

[51] P. Wesseling. *An introduction to multigrid methods*. John Wiley & Sons, Chichester, 1992.

[52] P. Wesseling. *Numerical solution of the stationary Navier-Stokes equations by means of a multiple grid method and Newton iteration*. Technical Report Report NA-18, Dept. of Technical Math. and Inf., Delft University of Technology, 1977.

[53] P. Wesseling. A robust and efficient multigrid method. In *Hackbusch and Trottenberg (1982)*, pages 614–630.

[54] P. Wesseling. Two remarks on multigrid methods. In *Hackbusch (1989)*, pages 209–216.

[55] P. Wesseling and P. Sonneveld. Numerical experiments with a multiple grid and a preconditioned Lanczos type method. In R. Rautmann, editor, *Approximation methods for Navier-Stokes problems*, pages 543–562, Springer-Verlag, Berlin, 1980. Lecture Notes in Math. 771.

[56] G. Wittum. Linear iterations as smoothers in multigrid methods: theory with applications to incomplete decompositions. *Impact of Computing in Science and Engineering*, 1:180–215, 1989.

[57] G. Wittum. On the robustness of ILU smoothing. *SIAM J. Sci. Stat. Comp.*, 10:699–717, 1989.

Incomplete Line LU as smoother and as preconditioner

P.M. de Zeeuw

Centre for Mathematics and Computer Science

P.O. Box 4079, 1009 AB Amsterdam, The Netherlands

E-mail: pauldz@cwi.nl

Abstract

Results are reported for Incomplete line LU (ILLU) in two different roles. One role is the role of smoother in a multigrid method for the solution of linear systems resulting from the 9-point discretization of a general linear second-order elliptic PDE in two dimensions. Together with features like matrix-dependent gridtransferoperators we obtain a blackbox multigrid solver (MGD9V). Another role for ILLU is as preconditioner in a stabilised bi-cg method (Bi-CGSTAB), as recently developed by Van der Vorst. In this role the preconditioner can easily be generalized for a discretized system of PDEs. A comparison is made between the two different roles.

1   Introduction

We consider a general linear 2nd order elliptic PDE in two dimensions

$$-\nabla \cdot (D\nabla u) + b_1(x,y)\frac{\partial u}{\partial x} + b_2(x,y)\frac{\partial u}{\partial y} + c(x,y)u = f(x,y) \tag{1}$$

on a bounded domain. $D(x,y)$ is a positive definite 2×2 matrix function and $c(x,y) \geq 0$. The discretized equation can be solved both by conjugate gradient and multigrid methods. Already for a long time incomplete decompositions have been applied fruitfully within these type of methods. In section 2 a description is given of the specific incomplete decomposition that we consider in this paper. The use of ILLU as smoother in multigrid methods is considered in section 3. Mere application of ILLU is not sufficient to ensure robustness and matrix-dependent gridtransfer operators are needed as well. Recently Van der Vorst developed the Bi-CGSTAB method [11] which is a variant of the method of Induced Dimension Reduction (IDR) as developed by Sonneveld [14]. We consider the use of ILLU as preconditioner for this method in section 4 and make a comparison with multigrid. In section 5 we generalize ILLU for the case of discretized systems of PDEs and we consider the application within Bi-CGSTAB. We end up with concluding remarks in section 6.

2   Incomplete line LU

The incomplete line LU decomposition (ILLU) has been originated by Underwood [10], and has also been proposed and elaborated upon by Concus, Golub and Meurant [4], Axelsson [2, 3], Meijerink [8] and others. In [6, 9] an extensive description of the method can be found. Here we repeat the general outline of the method. We assume to have a discretization on a rectangular computational grid that may be curvilinear in the geometrical sense. Let n_x denote the number of volumes in the x-direction in the case of a cell-centered discretization or the number of vertical lines in the case of a vertex-centered discretization. Likewise we define n_y, corresponding with the y-direction. Further we assume the common five point coupling (as with central differences) or nine point coupling (as with bilinear finite elements). With these assumptions we obtain after discretization a block tridiagonal linear system of the form

$$Ax = b \tag{2}$$

where

$$A = \begin{pmatrix} D_1 & U_1 & & & & \\ L_2 & D_2 & U_2 & & & \\ & L_3 & D_3 & \cdot & & \\ & & \cdot & \cdot & \cdot & \\ & & & \cdot & \cdot & \cdot \\ & & & & \cdot & D_{n_y} \end{pmatrix}. \tag{3}$$

The block D_j has the tridiagonal form:

$$D_j = \begin{pmatrix} d_{1j} & u_{1j} & & & & \\ l_{2j} & d_{2j} & u_{2j} & & & \\ & l_{3j} & d_{3j} & \cdot & & \\ & & \cdot & \cdot & \cdot & \\ & & & \cdot & \cdot & \cdot \\ & & & & \cdot & d_{n_x j} \end{pmatrix}. \tag{4}$$

The blocks L_j and U_j are of dimension n_x, just like D_j. In case of five point stencils the blocks L_j and U_j are diagonal-matrices. In case of nine point stencils these blocks are tridiagonal. The point of the ILLU-method is to make an incomplete factorization of A by the following formulae:

$$\overline{D}_1 = D_1, \tag{5}$$
$$\overline{D}_j = D_j - \mathbf{tridiag}(L_j \overline{D}_{j-1}^{-1} U_{j-1}), \quad j = 2(1)n_y. \tag{6}$$

The operator $\mathbf{tridiag}()$ forces a block (by clipping) into the sparsity pattern of the D_j. Without this particular operator, the factorization of A would be a complete one.

Performing one ILLU-relaxation sweep requires the following steps:

ILLU-sweep:
$$r = b - Ax;$$
$$z_1 = r_1;$$
$$z_j = r_j - L_j \overline{D}_{j-1}^{-1} z_{j-1}, \quad j = 2(1)n_y;$$
$$c_{n_y} = \overline{D}_{n_y}^{-1} z_{n_y};$$
$$c_j = \overline{D}_j^{-1}(z_j - U_j c_{j+1}), \quad j = n_y - 1(-1)1;$$
$$x_{new} = x + c;$$

3 ILLU and multigrid

We adhere to the assumptions about the grid and the discretization as made in the previous section. The general concept of multigrid methods is assumed to be known. We have a set of increasingly coarser grids:

$$\Omega_l, \Omega_{l-1}, \ldots, \Omega_k, \ldots, \Omega_1.$$

The discretization on the finest grid Ω_l evokes the linear system

$$A_l u_l = f_l . \tag{7}$$

We have to define our specific choice for the prolongation operator P_k, the restriction operator R_{k-1} and the coarse grid matrices A_{k-1} $(k = 2, \ldots, l)$. For the restriction we choose

$$R_{k-1} = P_k^T. \tag{8}$$

Further we choose the Galerkin approximation

$$A_{k-1} = R_{k-1} A_k P_k. \tag{9}$$

Hence, once P_k has been chosen, R_{k-1} and A_{k-1} follow automatically. Definition (8) is an essential ingredient for a blackbox algorithm because now a user only needs to define his problem on the finest grid (for a discussion on the concept of multigrid blackbox solvers see [13]). We may consider possible choices for the prolongation operator. A standard choice is bilinear interpolation. This amounts to taking an equal average of values of the solution at neighbouring coarse gridpoints (see Figure 1).

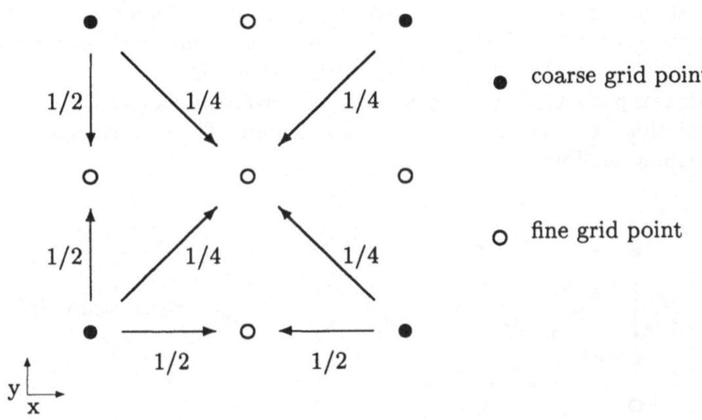

Figure 1: Bilinear prolongation.

At the gridpoints of the fine grid that coincide with the coarse grid we take identical values. One algorithm incorporating the foregoing prolongation is MGSYM [16]. It is a sawtooth multigrid correction scheme [12] which is furnished with ILLU as relaxation method. This algorithm works fine for a considerable class of problems. Yet the multigrid rate of convergence deteriorates severely at two different instances:

1. The diffusion coefficients in $D(x, y)$ are discontinuous across certain interfaces between subdomains.

2. The convection term is dominating, roughly speaking $h\|b\| > \|D\|$ with h the meshsize.

For an elaboration on the first instance we refer to [1, 7] and also to [5, § 10.3] where a one dimensional interface problem is analysed. A first example that the second instance causes divergence can be found in [17]. A complete analysis for this instance and a remedy can be found in [16], here the results of the analysis are illustrated when applied to the following simple example.

3.1 Example of Galerkin approximation for a convection dominated problem

Consider the following linear operator:

$$Au \equiv -\epsilon \Delta u + \frac{\partial u}{\partial x} \; . \tag{10}$$

For vanishing diffusion the following stencil is the result from a simple upwind discretization on a grid with meshsize equal to 1:

$$A_l \sim \begin{bmatrix} 0 & 0 & 0 \\ -1 & +1 & 0 \\ 0 & 0 & 0 \end{bmatrix}. \tag{11}$$

By repeatedly applying (8) and (9) for the standard choice we obtain on the n times coarsened grid Ω_{l-n} the stencil:

$$A_{l-n} \sim \begin{bmatrix} -\frac{1}{12} & \frac{1}{6} & -\frac{1}{12} \\ -\frac{1}{3} & \frac{2}{3} & -\frac{1}{3} \\ -\frac{1}{12} & \frac{1}{6} & -\frac{1}{12} \end{bmatrix} + 2^n \begin{bmatrix} -\frac{1}{12} & 0 & +\frac{1}{12} \\ -\frac{1}{3} & 0 & +\frac{1}{3} \\ -\frac{1}{12} & 0 & +\frac{1}{12} \end{bmatrix} + 2^{-n} \cdots \tag{12}$$

The two stencils are the same ones as evoked by discretization with bilinear finite elements of a diffusion and convection term in the $x-$direction. On the right hand side there is a remainder that decreases exponentially with n. What matters is the observation that the convection-stencil increases exponentially with n. The spurious solutions thus created on the coarse grids will affect severely the convergence of the multigrid algorithm as a whole.

A remedy for this particular example is to use an upwind prolongation instead, meaning here that only information from the left hand side is accepted. This corresponds to a prolongation with biassed weights, see Figure 2.

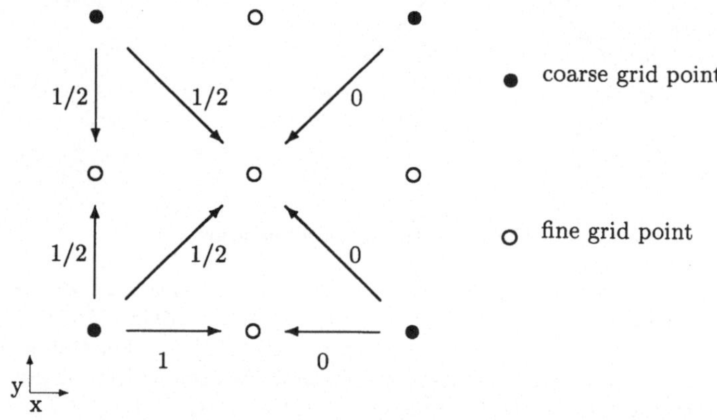

Figure 2: Example of upwind prolongation.

Now, by repeatedly applying (8) and (9) we obtain on the n times coarsened grid Ω_{l-n} the stencil:

$$A_{l-n} \sim 2^n \begin{bmatrix} 0 & 0 & 0 \\ -1 & +1 & 0 \\ 0 & 0 & 0 \end{bmatrix} + 2^{-n} \cdots \tag{13}$$

and the multigrid algorithm as a whole converges.

3.2 Matrix-dependent prolongations and restrictions

In [16] a prolongation operator has been proposed, able to handle both the case of dominant convection (in general directions) and interface problems at the same time. Here we give only an outline of the method. The grid Ω_k is split into four disjunct subgrids as follows:

$$\Omega_{k,(0,0)} \equiv \Omega_{k-1},$$

$$\Omega_{k,(1,0)} \equiv \{(x + h_k, y) \in \Omega_k \mid (x, y) \in \Omega_{k-1}\},$$

$$\Omega_{k,(0,1)} \equiv \{(x, y + h_k) \in \Omega_k \mid (x, y) \in \Omega_{k-1}\},$$

$$\Omega_{k,(1,1)} \equiv \{(x + h_k, y + h_k) \in \Omega_k \mid (x, y) \in \Omega_{k-1}\},$$

where h_k is the meshsize of grid Ω_k.

1. Let $\xi \in \Omega_{k,(1,0)}$ or $\xi \in \Omega_{k,(0,1)}$ be a point where we have to interpolate a coarse grid correction. Decompose the matrix A_k in its symmetric and antisymmetric part. The symmetric part S_k is supposed to correspond with diffusion and the zeroth order term, the antisymmetric part T_k with convection.

2. Reconstruct the diffusion and zeroth order coefficients at ξ from S_k, and the convection coefficients from T_k.

3. Use expressions that define an optimal choice for each sample of a set of degenerated cases for A_k.

4. At the fine grid points in $\Omega_{k,(0,0)}$ we adopt the values on Ω_{k-1}.

5. At the fine grid points in $\Omega_{k,(1,1)}$ we solve the homogeneous equation to obtain the correction.

Applying the derived formulae to the particular example of section 3.1 we obtain the upwind prolongation in Figure 2. Also here we adhere to (8) and (9), though the implementation of the latter is far from trivial. The actual computation of the coarse grid matrices takes less work than the ILLU-decompositions. The above is employed in the code MGD9V (de Zeeuw), this code uses the sawtooth multigrid correction scheme [12] and ILLU for smoother. For a detailed motivation of the prolongation and a description of the code, together with numerical experiments to illustrate its good behaviour, see [16].

4 ILLU and Bi-CGSTAB

In this section we consider the use of ILLU as preconditioner in Bi-CGSTAB (Van der Vorst [11]). A numerical example and comparison with the multigrid method MGD9V is presented at the end of this section. We have the linear system

$$Ax = b. \tag{14}$$

We consider two versions of Bi-CGSTAB: one with preconditioning from the right and one with preconditioning from the left. The first version reads:

right-Bi-CGSTAB:
$$x_0 = \underline{0};$$
$$\text{to } \sigma \text{ do } RELAX(A, x_0, b);$$
$$r_0 = b - Ax_0;$$
$$\rho_0 = \alpha = \omega_0 = 1;$$
$$v_0 = p_0 = \underline{0};$$
$$\text{for } i = 1, 2, 3, \ldots$$
$$\quad \rho_i = (r_0, r_{i-1}); \beta = (\rho_i/\rho_{i-1})(\alpha/\omega_{i-1});$$
$$\quad p_i = r_{i-1} + \beta(p_{i-1} - \omega_{i-1}v_{i-1});$$
$$\quad y = \underline{0}; \text{ to } \sigma \text{ do } RELAX(A, y, p_i)$$

$$v_i = Ay$$
$$\alpha = \rho_i/(r_0, v_i);$$
$$s = r_{i-1} - \alpha v_i;$$
$$z = \underline{0}; \text{ to } \sigma \text{ do } RELAX(A, z, s)$$
$$t = Az$$
$$\omega_i = (t, s)/(t, t);$$
$$x_i = x_{i-1} + \alpha y + \omega_i z;$$
$$r_i = s - \omega_i t;$$
$$\text{end}$$

At the i-th sweep this scheme delivers some approximation x_i of the solution x of (14), and the corresponding residual r_i. This residual is called the *updated* residual for the way in which it is computed.

Note that compared with preconditioning from both sides, we have gained a degree of freedom, for we can choose $\sigma > 1$. When we apply ILLU for RELAX we have to compute $(2+2(\sigma-1))$ matrix times vector operations for each i. Therefore, choosing $\sigma = 2$ instead of $\sigma = 1$ roughly doubles the amount of work per sweep. Yet, various numerical experiments have indicated that $\sigma = 2$ provides a more efficient choice for this parameter because of the faster convergence. Still higher values of σ decrease the effiency.

Note further that the initial guess for x_0 is determined by performing σ ILLU-sweeps on the zero solution.

We now introduce a variant of Bi-CGSTAB, based on preconditioning from the left.

left-Bi-CGSTAB:

$$x_0 = \underline{0};$$
$$\text{to } \sigma \text{ do } RELAX(A, x_0, b);$$
$$r_0 = b - Ax_0;$$
$$\tilde{r}_0 = \underline{0}; \text{ to } \sigma \text{ do } RELAX(A, \tilde{r}_0, r_0);$$
$$\rho_0 = \alpha = \omega_0 = 1;$$
$$v_0 = p_0 = \underline{0};$$
$$\text{for } i = 1, 2, 3, \ldots$$
$$\quad \rho_i = (\tilde{r}_0, \tilde{r}_{i-1}); \beta = (\rho_i/\rho_{i-1})(\alpha/\omega_{i-1});$$
$$\quad p_i = \tilde{r}_{i-1} + \beta(p_{i-1} - \omega_{i-1}v_{i-1});$$
$$\quad v_i = \underline{0}; \text{ to } \sigma \text{ do } RELAX(A, v_i, Ap_i)$$
$$\quad \alpha = \rho_i/(\tilde{r}_0, v_i);$$
$$\quad s = \tilde{r}_{i-1} - \alpha v_i;$$
$$\quad t = \underline{0}; \text{ to } \sigma \text{ do } RELAX(A, t, As)$$
$$\quad \omega_i = (t, s)/(t, t);$$
$$\quad x_i = x_{i-1} + \alpha p_i + \omega_i s;$$
$$\quad \tilde{r}_i = s - \omega_i t;$$
$$\text{end}$$

At the i-th sweep this scheme delivers some approximation x_i of the solution x of (14). An important feature of this second scheme is that the gridfunction $\tilde{r}_i$ is *not* the updated residual belonging to system (14). Instead, it is the updated residual belonging to a left-preconditioned version of this system. The preconditioning consists of the approximate inverse of A corresponding to σ relaxation sweeps. An important advantage of this new version of Bi-CGSTAB is that $\tilde{r}_i$ is a properly *scaled* residual. In fact, when applying ILLU for RELAX, $\tilde{r}_i$ will be a close approximation of the *error* rather than the residual. This is of importance e.g. within the context of semiconductor problems. Jacobians originating from this problems depict entries that differ in orders of magnitude. This make it hard to decide whether the residual is small or not. For making this kind of decisions (criterions etc.) the $\tilde{r}_i$ as approximation of the error, is a more convenient tool (see [15]).

4.1 Problem 1

The following testproblem has been proposed by Van der Vorst [11]. It is a simplified aquifer-problem, the convection-diffusion equation reads

$$-\nabla \cdot (D\nabla u) + b(x,y)\frac{\partial u}{\partial x} = f(x,y)$$
$$\Omega = (0,1) \times (0,1)$$

where the diffusion coefficient function D can be read from Figure 2, and

$$b(x,y) = 2\exp(2(x^2 + y^2)).$$

We have Dirichlet boundary conditions: $u = 1$ on $\partial\Omega$ except for $y = 1$ where $u = 0$. The function $f(x,y)$ equals zero everywhere, except for the small (dashed) square in the centre where $f(x,y) = 100$. We use meshsize $h = 1/130$, leading to a system with 129^2 unknowns.

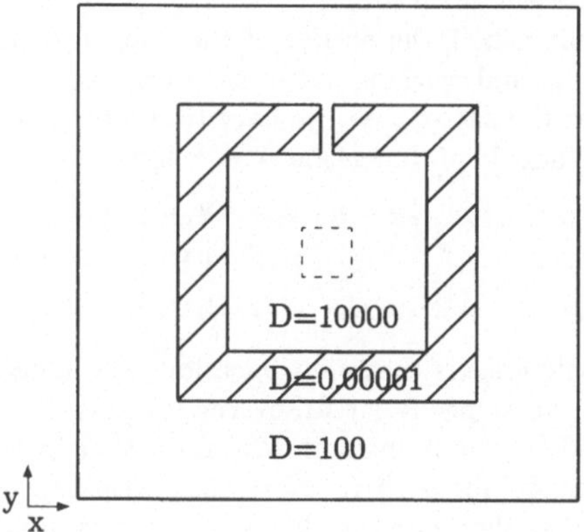

Figure 3: Geometry of Van der Vorst's aquifer-problem.

Here we use the same discretization as chosen by Van der Vorst, i.e. central differences. We observe that the diffusion coefficients are discontinuous across several interfaces and that at the shaded subdomain the convection is dominant.

In Figure 4 we see the 10-logarithm of the Euclidean norm of the iteration vectors r_i versus consumed workunits. One Bi-CGSTAB sweep ($\sigma = 2$) takes 2 workunits, one MGD9V cycle takes 1 workunit. For MGD9V and **right-Bi-CGSTAB** the r_i are residuals, as for **left-Bi-CGSTAB** the r_i are *scaled* residuals. We observe that from the viewpoint of efficiency, MGD9V has a clear advantage over the Bi-CGSTAB algorithms. Along with ILLU as relaxation method this is due to advanced features like matrix-dependent gridtransfers and an automatic Galerkin approximation of coarse grid matrices. However, it is far from trivial to generalize these features from the case of a scalar equation to the case of a system of coupled equations. As for Bi-CGSTAB, we show in the next section that, with a proper approach, the said generalization is rather straightforward.

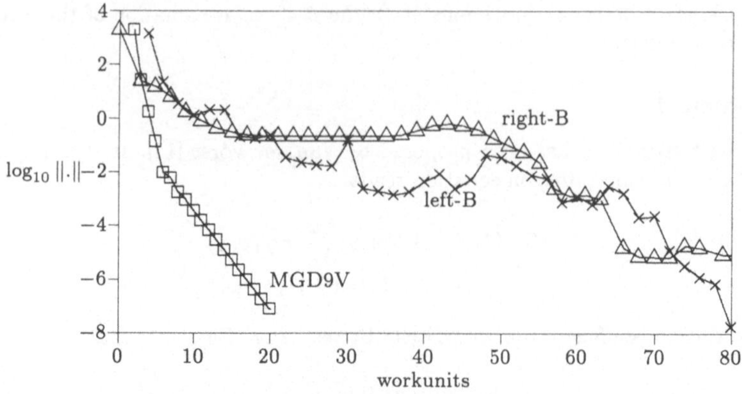

Figure 4: Convergence history of MGD9V and Bi-CGSTAB for the aquifer problem.

5 ILLU for discretized systems of PDEs

Suppose we have a system of n coupled PDEs. For $n = 1$ we obtain the matrix A as described in section 2. For $n > 1$ the entries of the matrix A become blocks of dimension n instead of scalars. At the actual working-out of the formulae for the decomposition, as known and described (e.g. [6]) for the scalar case, we have to replace operations on scalars x and y by operations on matrices X and Y of dimension n as follows:

$$x \pm y \longrightarrow X \pm Y$$
$$xy \longrightarrow XY$$
$$x/y \longrightarrow XY^{-1}.$$

However, an important difference is that multiplication is no longer commutative and therefore the working-out of the formulae needs careful overhauling.

In section 4 we mentioned the important advantage that in **left-Bi-CGSTAB** a properly scaled residual was computed for each i. Note that with the preconditioner for discretized coupled systems of PDEs, as developed in this section, we manage to do the scaling also over this coupling. For applications on semiconductor problems see [15].

6 Concluding remarks

We have given a description of a robust blackbox multigrid solver (MGD9V). Apart from ILLU as robust smoother, it features matrix-dependent gridtransferoperators in order to tackle problems with discontinuous diffusion and/or dominant convection. These features are hard to generalize to the case of a system of coupled PDEs. The solver (written in standard FORTRAN 77) is available from the author. We have also given descriptions of particular versions of the Bi-CGSTAB algorithm of Van der Vorst, furnished with ILLU as robust preconditioner. These versions have been generalized to the case of a system of PDEs. The left preconditioned version is especially suited for problems stemming from the field of semiconductor equations. This version might be of equal importance in other applications, e.g. fluid dynamics.

Acknowledgements

The author wishes to thank Prof. Dr. Henk A. Van der Vorst for providing his subroutine that discretizes the aquifer problem of section 4.1.

References

[1] R.E. Alcouffe, A. Brandt, J.E. Dendy Jr. and J.W. Painter, *The multi-grid method for the diffusion equation with strongly discontinuous coefficients*, SIAM J. Sci. Statist. Comput. **2**(4), pp. 430-454, 1981.

[2] O. Axelsson, *A General Incomplete Block-Matrix Factorization Method*, Linear Algebra and its Applications **74**, pp. 179-190, 1986.

[3] O. Axelsson, *A survey of preconditioned iterative methods for linear systems of algebraic equations*, BIT **25**, pp. 166-187, 1985.

[4] P. Concus, G.H. Golub and G. Meurant, *Block preconditioning for the conjugate gradient method*, SIAM J. Sci. Statist. Comput., **6** (1), pp. 220-252, 1985.

[5] W. Hackbusch, *Multi-Grid Methods and Applications*, Springer Ser. Comput. Math. **4**, Berlin, 1985.

[6] P.W. Hemker and P.M. de Zeeuw, *Some implementations of multigrid linear system solvers* in: D.J. Paddon and H. Holstein (Eds.) *Multigrid methods for integral and differential equations*, Inst. Math. Appl. Conf. Ser. New Ser. (Oxford Univ. Press, New York), pp. 85-116, 1985.

[7] R. Kettler, *Analysis and comparison of relaxation schemes in robust multigrid and preconditioned conjugate gradient methods*, in: W. Hackbusch and U. Trottenberg (Eds.), Lecture Notes in Math. **960** (Springer, Berlin), pp. 502-534, 1981.

[8] R. Kettler and J.A. Meijerink, *A multigrid method and a combined multigrid-conjugate gradient method for elliptic problems with strongly discontinuous coefficients in general domains*, Shell Publ. 604, KSEPL, Rijswijk, The Netherlands, 1981.

[9] P. Sonneveld, P. Wesseling and P.M. de Zeeuw, *Multigrid and conjugate gradient methods as convergence acceleration techniques*, in: D.J. Paddon and H. Holstein (Eds.) *Multigrid methods for integral and differential equations*, Inst. Math. Appl. Conf. Ser. New Ser. (Oxford Univ. Press, New York), pp. 117-167, 1985.

[10] R.R. Underwood, *An approximate factorization procedure based on the block Cholesky decomposition and its use with the conjugate gradient method*, Report NEDO-11386, General Electric Co., Nuclear Energy Div., San Jose, California, 1976.

[11] H.A. Van der Vorst, *Bi-CGSTAB: A fast and smoothly converging variant of Bi-CG for the solution of nonsymmetric linear systems*, SIAM J. Sci. Statist. Comput. **13**(2), 1992.

[12] P. Wesseling, *A robust and efficient multigrid method*, in: W. Hackbusch and U. Trottenberg (Eds.), Lecture Notes in Math. **960** (Springer, Berlin), pp. 614-630, 1981.

[13] P. Wesseling, *An Introduction to Multigrid Methods*, John Wiley & Sons Ltd, Chichester, England, 1991.

[14] P. Wesseling, P. Sonneveld, *Numerical experiments with a multiple grid and a preconditioned Lanczos type method*, in: R. Rautmann (ed.), *Approximation methods for Navier-Stokes problems*, Proceedings, Paderborn, Germany 1979, Lecture Notes in Mathematics, Springer, Berlin-etc., 1980.

[15] P.M. de Zeeuw, *Incomplete line LU for discretized coupled PDEs as preconditioner in Bi-CGSTAB*, Report, Centre for Mathematics and Computer Science, Amsterdam, to appear.

[16] P.M. de Zeeuw, *Matrix-dependent prolongations and restrictions in a blackbox multigrid solver*, Journal of Computational and Applied Mathematics, **33**, pp. 1-27, 1990.

[17] P.M. de Zeeuw and E.J. van Asselt, *The convergence rate of multi-level algorithms applied to the convection-diffusion equation*, SIAM J. Sci. Statist. Comput. **6**(2), pp. 492-503, 1985.

Bader, Dr. Georg Institut für Angewandte Mathematik, Universität Heidelberg, Im Neuenheimer Feld 294, W - 6900 Heidelberg

Barker, Prof. Vincent A. Institute for Numerical Analysis, The Technical University of Denmark, Building 305, DK - 2800 Lyngby, Denmark

Bastian, Dipl.-Inf. Peter IWR der Universität Heidelberg, Im Neuenheimer Feld 368, W - 6900 Heidelberg

Beauwens, Prof. Dr. R. Univ. Libre de Bruxelles, Faculté des Sciences Appliquées, 50, Av. F. D. Roosevelt, B - 1050 Bruxelles, Belgium

Benecke, Dipl.-Ing. Marcus Technische Fakultät der Universität Kiel, Lehrstuhl Prof. Diks, W - 2300 Kiel 1

Bey, Dipl.-Math. Jürgen Mathematisches Institut, Universität Tübingen, Auf der Morgenstelle 10, W - 7400 Tübingen

Braess, Prof. Dr. Dietrich Institut für Mathematik, Ruhr-Universität, W - 4630 Bochum 1

Broszeit, Dipl.-Math. Jens Institut für Strömungslehre und Strömungsmaschinen, Fachbereich Maschinenbau, Universität der Bundeswehr, Holstenhofweg 85, W - 2000 Hamburg 70

Burmeister, Jens Institut für Informatik und Praktische Mathematik, Christian-Albrechts-Universität Kiel, Olshausenstr. 40, W - 2300 Kiel 1

Dörfler, Willy Institut für Angewandte Mathematik der Universität Zürich, Rämistraße 74, CH - 8001 Zürich, Switzerland

Dorok, Dipl.-Math. Olfried Technische Universität "Otto von Guericke" Magdeburg, Institut für Analysis, Postfach 4120, O - 3010 Magdeburg

Dreyer, Dipl.-Math. Thomas Sonderforschungsbereich 123, Universität Heidelberg, Im Neuenheimer Feld 294, W - 6900 Heidelberg

Eiermann, M. Institut für Praktische Mathematik, Universität Karlsruhe, Englerstraße 2, W - 7500 Karlsruhe

Engels, Prof. Dr. Hermann Institut für Geometrie und Praktische Mathematik der RWTH Aachen, Templergraben 55, W - 5100 Aachen

Engl, Dipl.-Math. Gabriele Mathematisches Institut, Technische Universität München, Arcisstraße 21, Postfach 20 24 20, W - 8000 München

Gärtner, Dipl.-Math. R. Oldenburger Str. 2, W - 1000 Berlin 21

Gervang, Bo Dept. of Mathematics, UCW, Aberystwyth SY23 3BZ, Wales, UK

Graf, Kai JAFO-Technologie, Postfach 80 05 27, W - 2050 Hamburg 80

Gustavson, Dr. Ivar Institutionen för Informationsbehandling, Numerisk Analys, Chalmers Tekniska Högskola, S - 41296 Göteborg, Sweden

Haack, Dipl.-Ing. C. Technische Universität Hamburg-Harburg, Arbeitsbereich Meerestechnik II, Eißendorfer Str. 42, W - 2100 Hamburg 90

Hackbusch, Prof. Dr. Wolfgang Institut für Informatik und Praktische Mathematik, Christian-Albrechts-Universität Kiel, Olshausenstr. 40, W - 2300 Kiel 1

Hietel, Dipl.-Math. Dietmar Technische Hochschule Darmstadt, Fachbereich Mathematik, AG Numerische Mathematik, Schloßgartenstr. 7, W - 6100 Darmstadt

Horton, Dr. Graham IMMD III der Universität Erlangen-Nürnberg, Martensstr. 3, W - 8520 Erlangen

Johnson, Prof. Dr. Claes Chalmers University of Technology and University of Göteborg, Department of Mathematics, S - 41296 Göteborg, Sweden

Junkherr, Jörg Institut für Informatik und Praktische Mathematik, Christian-Albrechts-Universität Kiel, Olshausenstr. 40, W - 2300 Kiel 1

Kaaschieter, Dr. E. F. Eindhoven University of Technology, Department of Mathematics and Computer Science, P.O. Box 513, NL - 5600 MB Eindhoven, The Netherlands

Kapust, Birgit Institut für Informatik und Praktische Mathematik, Christian-Albrechts-Universität Kiel, Olshausenstr. 40, W - 2300 Kiel 1

Katzer, Dr. Edgar Institut für Informatik und Praktische Mathematik, Christian-Albrechts-Universität Kiel, Olshausenstr. 40, W - 2300 Kiel 1

Klaas, Ottmar Dorotheenstr. 7, Zimmer 695, W - 3000 Hannover 21

Knirsch, Ralf Institut für Rechnerstrukturen, Martensstr. 3, W - 8520 Erlangen

Kolotilina, L. Yu. The Leningrad branch of the V. A. Steklov Mathematical Institute, Academy of Sciences, Fontanka 27, 191011 Leningrad, Russia

Kornhuber, Dr. Ralf Konrad Zuse Zentrum Berlin, Heilbronner Str. 10, W - 1000 Berlin 31

Kreienmeyer, Dipl.-Math. Monika Universität Hannover, Institut für Baumechanik und Numerische Mechanik, Appelstraße 9 A, W - 3000 Hannover 1

Langer, Prof. Dr. Ulrich Technische Universität Chemnitz, Fachbereich Mathematik, Postfach 964, O - 9010 Chemnitz

Liebau, Frank Institut für Informatik und Praktische Mathematik, Christian-Albrechts-Universität Kiel, Olshausenstr. 40, W - 2300 Kiel 1

Lindskog, Dr. Gunhild Department of Computer Science, Chalmers University of Technology, S - 41296 Göteborg, Sweden

monga-Made, Dr. Magolu Université Libre de Bruxelles, Faculté des Sciences Appliquées, 50, Av. F. D. Roosevelt, B - 1050 Bruxelles, Belgium

Mahrenholtz, Prof. Dr. Ing. O. Arbeitsbereich Meerestechnik II, Technische Universität Hamburg-Harburg, Eißendorfer Str. 38, W - 2000 Hamburg-Harburg

Martensen, Prof. Dr. E. Mathematisches Institut II, Universität Karlsruhe (TH), Englerstraße 2, W - 7500 Karlsruhe

Mehrmann, Dr. Volker Universität Bielefeld, Fakultät für Mathematik, Postfach 8640, W - 4800 Bielefeld 1

Molnarka, Prof. Dr. G. Eötvös Lorand University, Department of Numerical Analysis, H - 1117 Budapest Bogdanfy v. 10/B, Hungary

Mordhorst, Uwe Rechenzentrum der Universität Kiel, Olshausenstr. 40 - 60, W - 2300 Kiel 1

Neuß, Nicolas IWR der Universität Heidelberg, Im Neuenheimer Feld 368, W - 6900 Heidelberg

Notay, Dr. Y. Metrologie Nucléaire, ULB, 50, av. F. D. Roosevelt, B - 1050 Bruxelles, Belgium

Pfau, Prof. Dr. Ing. H. Sektion Mathematik, Technische Hochschule, Juri-Gagarin-Ring 16, O - 2400 Wismar

Ploeg, Dr. Auke van der University of Groningen, Department of Mathematics, P.O. Box 800, NL - 9700 AV Groningen

Popa, Dr. Constantin Department of Mathematics, University of Constanta, Bd. Mamaia, Nr. 124, Constanza 8700, Romania

Rannacher, Prof. Dr. R. Institut für Angewandte Mathematik, Universität Heidelberg, Im Neuenheimer Feld 294/293, W - 6900 Heidelberg

Reichert, Henrik IWR der Universität Heidelberg, Im Neuenheimer Feld 368, W - 6900 Heidelberg

Reusken, Dr. Arnold TU Eindhoven, Fac. Wiskunde en Informatica, Postbus 513, NL - 5600 MB Eindhoven, The Netherlands

Rubart, Dipl.-Math. Lothar Institut für Strömungslehre und Strömungsmaschinen, Fachbereich Maschinenbau, Universität der Bundeswehr, Holstenhofweg 85, W - 2000 Hamburg 70

Sauter, Stefan Institut für Informatik und Praktische Mathematik, Christian-Albrechts-Universität Kiel, Olshausenstr. 40, W - 2300 Kiel 1

Schlegel, Dr. Ing. V. Arbeitsbereich Meerestechnik II, Technische Universität Hamburg-Harburg, Eißendorfer Str. 42, W - 2100 Hamburg 90

Schlüter, Dr. H.-J. Universität - GH Duisburg, FB 7 / Schiffstechnik, Lotharstraße, W - 4100 Duisburg 1

Schmidt, Thorsten Institut für Informatik und Praktische Mathematik, Christian-Albrechts-Universität Kiel, Olshausenstr. 40, W - 2300 Kiel 1

Schreck, Dipl.-Phys. Eberhard Lehrstuhl für Strömungsmechanik, Universität Erlangen-Nürnberg, Cauerstr. 4, W - 8520 Erlangen

Schulz, Heiko Mathematisches Institut A, Universität Stuttgart, Pfaffenwaldring 57, W - 7000 Stuttgart 80

Stevenson, Dr. R. P. Department of Mathematics, Budapestlaan 6, P.O. Box 80.010, NL - 3508 TA Utrecht, The Netherlands

Tan, Drs. K. Delft Hydraulics / dept S&O, P.O. Box 152, NL - 8300 Emmeloord, The Netherlands

Tröndle, Georg Institut für Angewandte Mechanik, Abt-Jerusalemstr. 7, W - 3300 Braunschweig

Turek, Dr. Stefan Sonderforschungsbereich 123, Universität Heidelberg, Im Neuenheimer Feld 294, 6900 Heidelberg

Vorst, Prof. Dr. H. A. van der Mathematisches Institut, Rijks Universiteit Utrecht, P.O. Box 80.010, NL - 3508 TA Utrecht, The Netherlands

Weiler, Dipl.-Math. Werner (†) Universität Heidelberg, Institut für Angewandte Mathematik, Sonderforschungsbereich 123, Im Neuenheimer Feld 294, 6900 Heidelberg

Wesseling, Prof. P.   Delft University of Technology, Faculty of Techn. Math. and Inform., Mekelweg 4, NL - 2628 CD Delft, The Netherlands

Wittekindt, Dipl.-Math. Jörg   Technische Hochschule Darmstadt, Fachbereich Mathematik, AG Numerische Mathematik, Schloßgartenstr. 7, W - 6100 Darmstadt

Wittum, Prof. Dr. Gabriel   IWR der Universität Heidelberg, Im Neuenheimer Feld 368, W - 6900 Heidelberg

Yeremin, Dr. A. Yu.   The Department of Numerical Mathematics of the Academy of Sciences, Ryleev Street 29, 119 034 Moscow, Russia

Yserentant, Prof. Dr. H.   Mathematisches Institut, Universität Tübingen, Auf der Morgenstelle 10, W - 7400 Tübingen

Zeeuw, P. M. de   Department of Numerical Mathematics (NW), Centre for Mathematics and Computer Science, P.O. Box 4079, NL - 1009 AB Amsterdam, The Netherlands

Zeng, Shi   Delft University of Technology, Faculty of Techn. Math. and Inform., Mekelweg 4, NL - 2628 CD Delft, The Netherlands

Notes on Numerical Fluid Mechanics (NNFM)   Volume 41

Series Editors: Ernst Heinrich Hirschel, München
Kozo Fujii, Tokyo
Bram van Leer, Ann Arbor
Keith William Morton, Oxford
Maurizio Pandolfi, Torino
Arthur Rizzi, Stockholm
Bernard Roux, Marseille

Addresses of the Editors of the Series "Notes on Numerical Fluid Mechanics"

Prof. Dr. Ernst Heinrich Hirschel (General Editor)
Herzog-Heinrich-Weg 6
D-8011 Zorneding
Federal Republic of Germany

Prof. Dr. Kozo Fujii
High-Speed Aerodynamics Div.
The ISAS
Yoshinodai 3-1-1, Sagamihara
Kanagawa 229
Japan

Prof. Dr. Bram van Leer
Department of Aerospace Engineering
The University of Michigan
Ann Arbor, MI 48109-2140
USA

Prof. Dr. Keith William Morton
Oxford University Computing Laboratory
Numerical Analysis Group
8-11 Keble Road
Oxford OX1 3QD
Great Britain

Prof. Dr. Maurizio Pandolfi
Dipartimento di Ingegneria Aeronautica e Spaziale
Politecnico di Torino
Corso Duca Degli Abruzzi, 24
I-10129 Torino
Italy

Prof. Dr. Arthur Rizzi
FFA Stockholm
Box 11021
S-16111 Bromma II
Sweden

Dr. Bernard Roux
Institut de Mécanique des Fluides
Laboratoire Associé au C.R.N.S. LA 03
1, Rue Honnorat
F-13003 Marseille
France

Brief Instruction for Authors

Manuscripts should have well over 100 pages. As they will be reproduced photomechanically they should be typed with utmost care on special stationary which will be supplied on request.
In print, the size will be reduced linearly to approximately 75 per cent. Figures and diagrams should be lettered accordingly so as to produce letters not smaller than 2 mm in print. The same is valid for handwritten formulae. Manuscripts (in English) or proposals should be sent to the general editor, Prof. Dr. E. H. Hirschel, Herzog-Heinrich-Weg 6, D-8011 Zorneding.